Rebecca Saive

Potential Distribution within Organic Solar Cells

Rebecca Saive

Potential Distribution within Organic Solar Cells

Investigation of the Potential Distribution within Organic Solar Cells by Scanning Kelvin Probe Microscopy

Südwestdeutscher Verlag für Hochschulschriften

Impressum / Imprint
Bibliografische Information der Deutschen Nationalbibliothek: Die Deutsche Nationalbibliothek verzeichnet diese Publikation in der Deutschen Nationalbibliografie; detaillierte bibliografische Daten sind im Internet über http://dnb.d-nb.de abrufbar.
Alle in diesem Buch genannten Marken und Produktnamen unterliegen warenzeichen-, marken- oder patentrechtlichem Schutz bzw. sind Warenzeichen oder eingetragene Warenzeichen der jeweiligen Inhaber. Die Wiedergabe von Marken, Produktnamen, Gebrauchsnamen, Handelsnamen, Warenbezeichnungen u.s.w. in diesem Werk berechtigt auch ohne besondere Kennzeichnung nicht zu der Annahme, dass solche Namen im Sinne der Warenzeichen- und Markenschutzgesetzgebung als frei zu betrachten wären und daher von jedermann benutzt werden dürften.

Bibliographic information published by the Deutsche Nationalbibliothek: The Deutsche Nationalbibliothek lists this publication in the Deutsche Nationalbibliografie; detailed bibliographic data are available in the Internet at http://dnb.d-nb.de.
Any brand names and product names mentioned in this book are subject to trademark, brand or patent protection and are trademarks or registered trademarks of their respective holders. The use of brand names, product names, common names, trade names, product descriptions etc. even without a particular marking in this work is in no way to be construed to mean that such names may be regarded as unrestricted in respect of trademark and brand protection legislation and could thus be used by anyone.

Coverbild / Cover image: www.ingimage.com

Verlag / Publisher:
Südwestdeutscher Verlag für Hochschulschriften
ist ein Imprint der / is a trademark of
OmniScriptum GmbH & Co. KG
Heinrich-Böcking-Str. 6-8, 66121 Saarbrücken, Deutschland / Germany
Email: info@svh-verlag.de

Herstellung: siehe letzte Seite /
Printed at: see last page
ISBN: 978-3-8381-3946-3

Zugl. / Approved by: Heidelberg, University, Diss., 2014

Copyright © 2014 OmniScriptum GmbH & Co. KG
Alle Rechte vorbehalten. / All rights reserved. Saarbrücken 2014

Untersuchung des Ladungstransports in organischen Halbleitern mittels Kelvin Rastersondenmikroskopie — Im Rahmen dieser Doktorarbeit wurde der Potentialverlauf innerhalb organischer Solarzellen untersucht. Dazu wurde in die Zellen ein wenige Mikrometer großes Loch mit einem fokussierten Ionenstrahl gefräst, so dass der Querschnitt der Zellen für Raster-Kelvin Mikroskopie (SKPM) zugänglich wurde. SKPM Messungen unter Beleuchtung und unter angelegter elektrischer Spannung wurden mit diesem Verfahren an Solarzellen aus Poly(3-hexylthiophen) (P3HT) und [6,6]-phenyl-C61-butyric acid methyl ester (PCBM) durchgeführt. In Solarzellen, in denen die Materialien in einer Zweischichtstruktur aufgebracht sind, fällt eine angelegte Spannung zwischen P3HT und Anode und über der gesamten organischen Schicht ab. In einer Zelle mit interpenetrierendem Schichtsystem (bulk heterojunction) findet der Potentialabfall bei angelegter Spannung am Kontakt zur Anode und am Kontakt zur Kathode statt. Invertiert man die Solarzellen durch Änderung der Kontaktmaterialien, so kann kein Potentialabfall an den Kontakten beobachtet werden, sondern das gesamte Potential fällt über der organischen Schicht ab. Daraus lässt sich schließen, dass invertierte Kontakte für diese Morphologie des interpenetrierenden Systems bevorzugt sind. Es konnte weiterhin gezeigt werden, dass die offene Klemmspannung am selben Ort abfällt wie eine von außen angelegte Spannung. SKPM Messungen wurden zudem an Solarzellen mit S-förmigen Strom-Spannungs-Kennlinien vorgenommen. Es konnte direkt abgebildet werden, dass die besagte S-Form aus einer Transportbarriere am Kontakt zur Kathode resultiert.

Investigation of charge transport in organic semiconductors by scanning Kelvin probe microscopy — In this work the potential distribution within organic solar cells was investigated. Using a focused ion beam micrometer sized holes were milled into the cells such that the cross sections became accessible by scanning Kelvin probe microscopy (SKPM). SKPM measurements were performed on Poly(3-hexylthiophen) (P3HT) and [6,6]-phenyl-C61-butyric acid methyl ester (PCBM) solar cells under illumination and under different bias voltages. In a bilayer solar cell the applied bias voltage drops at the interface between P3HT and anode and within the organic layer. In a bulk heterojunction solar cell the potential drops at the interface between P3HT and the anode and at the interface between the PCBM and cathode. In solar cells which were inverted due to altered contact materials there is no potential drop at the contacts, but the potential uniformly drops within the organic material. It can be concluded that an inverted device structure is more favorable for this morphology of the bulk heterojunction. The open circuit voltage exhibited a similar distribution within the device as an external applied bias voltage. Furthermore, SKPM measurements were performed on solar cells with S-shaped current-voltage characteristics. It was mapped that the S-shape behavior results from a transport barrier at the cathode interface.

Contents

1. **Introduction**   1

2. **Fundamentals of organic semiconductors**   5
   2.1. Charge transport in organic materials 5
   2.2. Interfaces 6
        2.2.1. Ideal interfaces 7
        2.2.2. Real interfaces 7
   2.3. Organic solar cells 8

3. **Preparation of samples**   13
   3.1. TIPS-pentacene OFETs 13
   3.2. P3HT:PCBM BHJ solar cells 13
   3.3. P3HT:PCBM bilayer solar cells 15

4. **Measurement methods and experimental setup**   17
   4.1. Kelvin probe 17
   4.2. Scanning probe microscopy 18
   4.3. Scanning Kelvin probe microscopy 19
   4.4. Electron microscopy and focused ion beam 22
   4.5. Experimental setup 24

5. **Evaluation of the method**   29

6. **SKPM on OFETs and Si solar cells**   35
   6.1. TIPS-pentacene OFET 35
   6.2. Cleaved Silicon solar cells 37

7. **Potential distribution within P3HT/PCBM BHJ and bilayer solar cells**   41
   7.1. P3HT/PCBM bilayer solar cells 41
        7.1.1. IV curves 41
        7.1.2. Analysis of the cross section 42
        7.1.3. SKPM data with applied bias voltage in the dark . . . 46

Contents

      7.1.4. SKPM data with applied bias voltage under illumination .. 49
      7.1.5. SKPM data in open circuit condition 49
  7.2. Conventional P3HT:PCBM BHJ solar cells 55
      7.2.1. IV curves 55
      7.2.2. SKPM data with applied bias voltage in the dark ... 56
      7.2.3. SKPM data in open circuit condition 56
  7.3. Inverted P3HT:PCBM BHJ solar cells 60
      7.3.1. IV curves 61
      7.3.2. SKPM data with applied bias voltage in the dark ... 62
      7.3.3. SKPM data in open circuit condition 62
  7.4. Discussion 62

8. S-shaped solar cells   **69**
  8.1. S-shaped P3HT/PCBM bilayer solar cells 70
      8.1.1. Results 70
      8.1.2. Discussion 72
  8.2. S-shaped P3HT:PCBM BHJ solar cell 76
      8.2.1. Results 76
      8.2.2. Discussion 78

9. Potential distribution within OLEDs   **81**

10. Summary and outlook   **85**

Bibliography   **89**

A. Appendix   **107**
  A.1. Notes on the measurement procedure 107
      A.1.1. Mounting of cantilever and sample 107
      A.1.2. Strange behavior of SKPM signal 108
      A.1.3. Drift due to heating 108

B. Publications   **109**

Abbreviation

AC:	Alternating current
AFM:	Atomic force microscopy
Alq$_3$:	8-tris-hydroxyquinoline aluminium
BHJ:	Bulk hetero junction
CPD:	Contact potential difference
CV:	Capacitance-voltage
DC:	Direct current
FIB:	Focused ion beam
HOMO:	Highest occupied molecular orbital
iL:	InnovationLab
ITO:	Indium-tin oxide
IV:	Current-voltage
KP:	Kelvin probe
LED:	Light emitting diode
LUMO:	Lowest unoccupied molecular orbital
NPB:	N,N´-di-(1-naphthyl)-N,N´-diphenyl-1,1´-biphenyl-4,4´-diamine
OFET:	Organic field effect transistor
OLED:	Organic light emitting diode
OSC:	Organic solar cell
P3HT:	Poly(3-hexylthiophen)

Contents

PCBM: [6,6]-phenyl-C61-butyric acid methyl ester

PEDOT:PSS: Poly(3,4-ethylenedioxythiophene):poly-(styrenesulfonate)

PEIE: Polyethylenimine ethoxylated

SEM: Scanning electron microscopy

SKPM: Scanning Kelvin probe microscopy

SPM: Scanning probe microscopy

STM: Scanning tunneling microscopy

TEM: Transmission electron microscopy

TIPS-pentacene: 6,13-bis(triisopropylsilylethynyl) pentacene

WF: Work function

1. Introduction

Energy supply is one of the hugest challenges modern society has to face. The growth of the world population and economic wealth lead to steadily increasing energy demand. By now, the world energy consumption amounts to 17 TW [1], the U.S. Energy Information Administration forecasts the demand to grow by about 56% to 27 TW by 2040 [1]. On the other hand, fossil fuel resources decrease or will be difficult to access. Furthermore, the burning of fossil resources releases bound carbon which has resulted in a significant increase of the carbon dioxide concentration in the atmosphere since the industrial revolution [2]. As carbon dioxide is a polar, infrared active molecule it can contribute to the greenhouse effect [3], so that many scientists see a correlation between carbon dioxide in the atmosphere and global warming [4].

Consequently, a lot of effort is made to reduce the irreversible consumption of fossil fuels and to develop sustainable energy supply. So called renewable energy sources are manifold. With a total power of about 10^{17} W [5] which reaches the earth on average, the sun provides a multiple of the energy we need. Wind energy, biomass and hydro energy (which also have their origin in sun radiation) are further renewable energy sources [6–8].

But the term "sustainable" does not only mean ecological aspects. Energy has to be affordable to ensure economic progress and social wealth. Nowadays, renewable energy sources are criticized for their high prduction cost (in Germany: 0.078-0.142 €/kWh for electrical energy [9]) compared to conventional energy sources like lignite (in Germany: 0,038-0.053 €/kWh for electrical energy [9]) or nuclear power. Many national administrations including the German Federal Government have addressed this issue by supporting the power production from renewables through offtake and price guarantees [10]. Considering only this aspect, renewable energy is not yet profitable which unfortunately is usually the only aspect covered by German media. On the other side of the coin, political influence on the energy sector is not an unusual thing and has also been active in favour of coal or nuclear energy in the past [11]. Furthermore, the enormous economical risks imposed by the climate change or the use and disposal of nuclear material are incalculable. Nevertheless, the ambition of researchers and industrials in the renewable energy sector is to narrow the gap of immediate production costs in comparison to conventional

1. Introduction

sources.

Sun power can be used in different ways. In solar heat power plants for example the sun heats molten salt [12, 13] or oil [14] which than heats water and drives turbines. A very elegant way to directly transform solar energy into electrical energy is provided by photovoltaic (PV) devices, also called solar cells. The photoelectric effect, first described as a quantum effect by Albert Einstein [15, 16], is used to convert the energy of light (photons) into electrical power by exciting electrons within the PV device [17]. PV devices consist of semiconducting materials. Silicon is the most prominent one from which the first semiconductor solar cell modul was made [18–20] and which is usually used in roof top PV power plants. But many other material systems like GaAs, CuInSe or CdTe are also used for PV and they all come with advantages and disadvantages. However, all commercially applied solar cells produce by far more energy during their live time than they consume during their production. The energy pay back time (period which is needed to produce the production energy) amounts to a few years depending on the type of solar cell and the region where it is installed. For an amorphous silicon solar cell in Germany the energy pay back time amounts to about two years [21].

A relatively new class of PV devices is formed by organic solar cells in which organic (carbon based) molecules serve as semiconducting materials. Since the discovery of semiconducting properties in organic molecules in 1977 by Alan Heeger [22], the investigation and application of organic semiconductors has boomed. They offer the possibility to realize flexible, low-cost and sustainable electronics. Smartphones and televisions with organic light emitting diode (OLED) displays already have been commercially fabricated for some years. First organic solar cells are also commercially available and they offer applications one cannot realize with inorganic solar cells. Organic solar cells are lightweight and can be coated on flexible substrates, such that they are very suitable for mobile applications like outdoor equipment. Furthermore, flexible substrates offer the opportunity of roll-to-roll fabrication [23]. As organic solar cells can be processed from solution, it is even possible to print organic devices [24, 25]. Organic solar cells can be fabricated in all colors and even transparent solar cells are possible [26, 27]. This property makes them interesting candidates for building integrated PV for example in windows. Furthermore, they have a very good absorbance of indirect light compared to crystalline inorganic semiconductors. Today, the world record efficiency of organic solar cells amounts to 10.7% [28]. This value increases steadily but compared to inorganic solar cells (silicon solar cell world record: 25.0% [28]) it is still low. However, the perspective of cheap production, alternative applications and the rapid progress justifies research efforts.

Although there is a huge progress in the performance of these devices, the physical fundamentals of charge transport are still lacking of a closed description. Weak bonding and disorder of the involved molecules often leads to properties strongly deviating from inorganic semiconductors. Many scientific groups have investigated the charge generation, separation and transport in organic solar cells by different methods like optical spectroscopy [29], scanning probe techniques [30–32] or nonlinear optical microscopy [33], but the potential distribution within operating solar cells and the origin of the open circuit voltage remain unclear. One well established method to investigate the surface potential was already used by many scientists to image the potential distribution in organic field effect transistors (OFETs) and thereby to gain information about the charge transport in these devices [34–37]. This method is called scanning Kelvin probe microscopy (SKPM). With SKPM, measurements of the contact potential difference can be performed spatially resolved and so the potential distribution within a device can be mapped. SKPM works similar to atomic force microscopy (AFM) which means that a tiny tip on a cantilever is scanned along the surface. Conventional SKPM is restricted to measure on the surface of devices whereas (contrary to OFETs) the charge transport in organic solar cells occurs vertically to the surface [38]. Therefore, one needs to expose the cross section of the solar cells to perform the potential mapping. In inorganic solar cells the cross section can be easily exposed due to their cristallinity by cleaving the samples. Cleaving of organic solar cells leads to rough surfaces which are difficult to acces by SKPM [39, 40].

Here, a new method will be presented in which the cross section is laid open by milling micrometer sized holes into the solar cells with a focused ion beam. Solar cells made from poly(3-hexylthiophen) (P3HT) and [6,6]-phenyl-C61-butyric acid methyl ester (PCBM) were investigated in bilayer, conventional bulk heterojunction (BHJ) and inverted BHJ device structures. SKPM measurements were performed on these devices in the dark, under illumination and with applied bias voltage. In a bilayer solar cell the applied bias voltage drops at the interface between P3HT and anode and within the organic layer. In a BHJ solar cell the potential drops at the interface between P3HT and the anode and at the interface between the PCBM and cathode. In solar cells which were inverted due to altered contact materials there is no potential drop at the contacts, but the potential uniformly drops within the organic material. It can be concluded, that an inverted device structure is more favorable for this morphology of the BHJ. The open circuit voltage exhibited a similar distribution within the device as an externally applied bias voltage. Furthermore, SKPM measurements were performed on solar cells with S-shaped current-voltage characteristics. For the first time, it could be

1. Introduction

directly mapped that the S-shape behavior results from a transport barrier at the cathode interface, which was already predicted by many other experiments and simulations.

This project was executed in the InnovationLab [41] in Heidelberg which is a joined research platform of industry and academia within the "Leading-Edge Cluster Forum Organic Electronic" sponsored by the Federal Ministry of Education and Research (BMBF). The institute covers the whole organic electronics' value chain from fundamental research to industrial application. The institute is divided into competence centers which work with different methods on different topics. The Simulations competence center for example uses different computer simulation methods to describe and predict properties of organic materials. The competence center Synthesis synthesizes new materials which are tested in the competence center Device Physics. The competence center Printing works out printing techniques from small inkjet printers up to a huge roll-to-toll printing machine. This dissertation project is affiliated to the competence center Analytics which investigates the fundamental properties of organic semiconductors. In a joined laboratory the Universities of Heidelberg, Braunschweig and Darmstadt bring together their competences in infrared spectroscopy, photoelectron spectroscopy, electron microscopy and scanning probe microscopy.

This dissertation is structured as follows: In chapter 2 an introduction into the fundamental properties of organic semiconductors will be given. Chapter 3 outlines the preparation of the investigated samples and in chapter 4 the measurement methods and the experimental setup will be explained. The above mentioned new method for the investigation of organic devices' cross sections by FIB milling and further analysis by SKPM will be evaluated in chapter 5. Chapter 6 will show measurements on an OFET and a silicon solar cell which were performed to reproduce literature results and to confirm that the launch of SKPM in the InnovationLab has been successful. Chapter 7 and chapter 8 constitute the main part of this dissertation as within these chapters the novel results on different P3HT/PCBM solar cells will be presented. Chapter 9 gives a short introduction into the results of SKPM measurements on cross sections of OLEDs. In chapter 10 a summary, conclusion and outlook will be given.

2. Fundamentals of organic semiconductors

In this chapter an introduction to the most important aspects of organic semiconductors will be given in order to understand the experimental results. It is important to know about the charge transport in organic materials as well as about the effects occurring at interfaces between different organic materials and at the metal contacts. Therefore, in the first two sections the charge transport in organic materials and at interfaces will be explained. In the third section organic solar cells will be introduced.

2.1. Charge transport in organic materials

Organic semiconductors consist of small organic molecules or polymers which for their part consist of monomer units.

If electrons can move between the molecules or the units the material in principle is able to conduct electrical current. The movement from one unit A to another unit B can be described by Fermi's Golden Rule [42–44]:

$$R_{A \to B} = \delta(E_B - E_A + \Delta E) \cdot \rho \cdot <A|M_{A \to B}|B>.$$

$R_{A \to B}$ is the transition rate from state A to state B. First of all, the energy conservation has to be obeyed. In the formula this is implemented by the Kronecker delta $\delta(E_B - E_A + \Delta E)$. If A and B are not on the same energy levels (E_A and E_B) or if there exists an energetic barrier between them, the electron has to be excited, e.g. by heating or by an electric field (ΔE). In case of similar energy levels with a barrier, electrons can overcome the barrier by tunneling. Second, there must be occupied states in the initial state and there must be vacant states in the final state. The density of states is considered by the factor ρ. The higher the state density the higher the transition rate. Third, there must be an overlap of the initial and the final state which is illustrated by the transition matrix element $<A|M_{A \to B}|B>$ in Fermis' Golden Rule. In this description the movement of the electrons usually is called hopping and a detailed introduction into hopping transport is given by H. Bässler [45].

The orbitals of the atoms in the molecules overlap and form molecular orbitals [46]. The molecular orbitals are filled by the electrons up to the highest occupied molecular orbital (HOMO). The lowest unoccupied molecular orbital (LUMO) lies energetically higher and it is possible to excite electrons from the HOMO to the LUMO such that an electron vacancy in the HOMO and an occupied state in the LUMO arise. The vacancy (hole) and the electron are able to participate in the hopping charge transport.

In the case of ordered organic crystals with sufficient orbital overlap Bloch's theorem [47, 48] becomes valid and one can find expanded, band like energy states where the overlapping HOMOs correspond to the valence band in a classic semiconductor and the LUMOs correspond to the conduction band [49].

Disorder or impurities broaden the bands and lead to tail and gap states [50, 51]. If the lattice becomes irregular the translation invariance vanishes [52, 53] and therefore quasi impulse conservation is no longer given [54, 55]. This is similar in amorphous silicon and e.g. leads to the good absorbance of indirect light [56, 57].

Materials with high charge carrier mobilities in the HOMO are called hole conductor or p-type, materials with high charge carrier mobilities in the LUMO are called electron conductor or n-type. Materials with high charge carrier mobilities in the HOMO and in the LUMO are called ambipolar.

Similar to inorganic semiconductors the charge transport in organic materials usually works better in ordered systems. Disorder and grain boundaries lead to charge transport barriers which have to be overcome by tunneling, by an electric field or by heating. Charge transport in an electronic device may be further hindered by mismatched contacts and interfaces.

2.2. Interfaces

An electronic device usually consists of different materials which have different functions in the device. E.g., in a simple organic solar cell, an organic hole and an electron conductor act as absorbers and separate the excitons. Metal contacts conduct the current to the consumer load. The interfaces between the organic materials as well as between the organic material and the metal play a significant role in the transport.

2.2. Interfaces

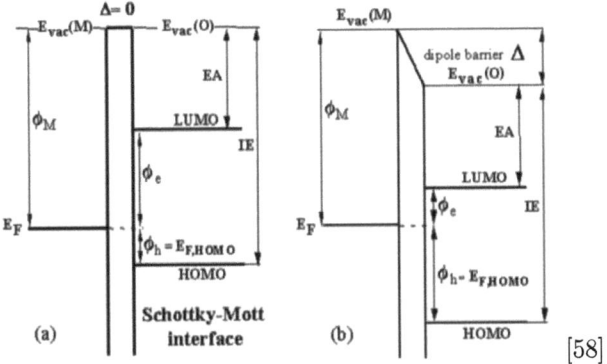

Figure 2.1: From [58]: "Energy diagram of a metal-organic semiconductor interface (a) without and (b) with a dipole barrier Δ. ϕ_e and ϕ_h are the electron and hole barriers, respectively, and $E_{vac}(O)$ and $E_{vac}(M)$ are the organic and metal vacuum levels, respectively".

2.2.1. Ideal interfaces

The Schottky-Mott model makes predictions on the band level alignment if a metal and a semiconductor get in contact. In this model a potential hill is formed at the interface which corresponds to the work function difference between the semiconductor and the metal [59, 60]. This potential hill shows rectifying behavior for n-doped semiconductors if the work function of the metal is higher than that of the semiconductor. If the work function of the metal is lower than that of the semiconductor, the interface acts as an ohmic contact. For p-doped semiconductors the interface shows rectifying behavior if the work function of the metal is lower than that of the semiconductor [61, 62]. The energy levels of inorganic as well as of organic materials can be measured by photoelectron spectroscopy [63].

2.2.2. Real interfaces

At real interfaces the Mott-Schottky model usually does not apply, but Fermi level pinning and the formation of interface dipoles [64–66] can occur. Fermi level pinning is a common phenomenon in inorganic semiconductors [67, 68] and organic semiconductors [69, 70]. Surface states lead to energy levels within the band gap which pin the Fermi level to a certain niveau such that the work function of the metal is no longer decisive for energy alignment [71]. Interface dipoles are held responsible for observed discontinuities in the energy

2. Fundamentals of organic semiconductors

levels at interfaces [72–74]. In figure 2.1 the energy alignment in a metal-organic semiconductor interface is schematically demonstrated. ϕ_e and ϕ_h are the electron and hole barriers, respectively, and $E_{vac}(O)$ and $E_{vac}(M)$ are the organic and metal vacuum levels, respectively. In figure 2.1a the energy alignment is shown by using the above explained Schottky-Mott model for thin organic layers with low charge carrier density. In figure 2.1b there exists a dipole barrier Δ between the metal and the organic semiconductor. Thus, the transport barrier for electrons and holes is altered.

In figure 2.2 an experiment by Antoine Kahn et al. is shown in which the energy alignment between different metals and eight different organic materials was investigated [58]. For each organic material the measured interface position of the Fermi level E_F with respect to HOMO and LUMO as a function of the metal work function is shown. The dashed oblique lines correspond to the Schottky–Mott limit of the Fermi level position, and the vertical lines give the magnitude of the measured interface dipole barriers. The predictions of the Schottky-Mott model are only valid in a few cases.

Furthermore, organic semiconductors chemically react with the contact materials which alters the injection properties [75–78].

Overall, one can state, that it is almost impossible to predict the properties of an interface from the energy levels of the single materials. Therefore, it is important to measure energy levels on interfaces or in realistic devices.

2.3. Organic solar cells

Solar cells convert the energy of light into electrical energy by absorbing photons which excites electrons from the valence band (or HOMO level) to the conduction band (or LUMO level) [17]. The electrons in the conduction band and the missing electrons in the valence band create electron-hole-pairs. They are bound by the Coulomb interaction and form a quasiparticle, the exciton. Due to the weak electrostatic shielding in organic semiconductors the bounding energy is in the range of 0.1-1.0 eV (Frenkel exciton [54, 80]) and thermal energy is not sufficient for their separation. Therefore most organic solar cells consist of a hole transporting material with a relatively high HOMO level and an electron transporting material with a relatively low LUMO level to separate the exciton at the interface between both materials. This is known as the charge transfer (CT) state and from this state free electrons and holes are formed [79]. The principles of photon absorption, exciton diffusion and dissociation are displayed in figure 2.3 in a spatial (a) and in an energetic (b) scheme. By passing the electron or hole conductor the separated charge

2.3. Organic solar cells

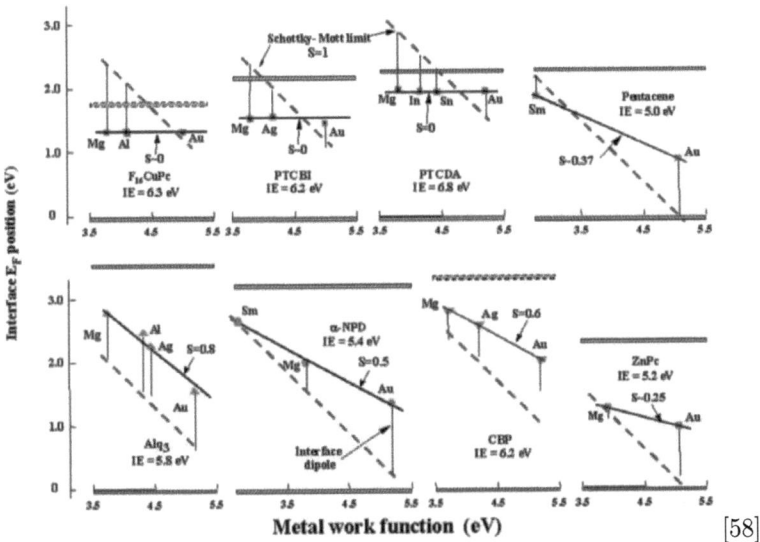

Figure 2.2: From [58]: "Measured interface position of E_F with respect to HOMO and LUMO as a function of the metal work function for eight different molecular materials. In each panel, the thick horizontal bottom and top bars represent the HOMO (with a work function scale) and LUMO, respectively. Dashed LUMO bars mean that the LUMO position is not precisely known. The data points were obtained via UPS for organic-on-metal interfaces. The dashed oblique lines correspond to the Schottky–Mott limit of the Fermi level position, and the vertical lines give the magnitude of the measured interface dipole barriers".

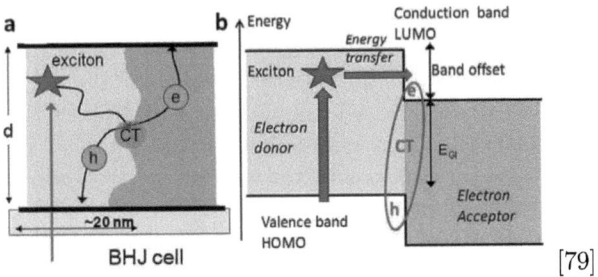

Figure 2.3: Principles of photon absorption, exciton diffusion and dissociation in a a) spatial and b) energetic scheme. From [79].

2. Fundamentals of organic semiconductors

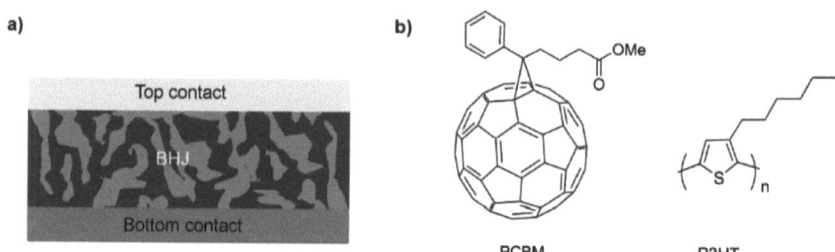

Figure 2.4: a) Scheme of a BHJ solar cell. b) Chemical structure of PCBM and P3HT. From [82].

carriers then arrive at the top and bottom contact where they are extracted if the energy levels at the interfaces match. The driving force for this charge carrier drift is given by an internal electric field which is built up by the work function difference of the contact materials [79]. There are also scientists who claim that the internal field is built up by the energy difference between the HOMO level of the donor and the LUMO level of the acceptor components [81]. Both hypotheses will be discussed within the experimental section.

The diffusion length of the excitons is in the range of a few nanometers so an efficient separation is only possible in thin layers. As the absorption becomes poor in thin layers the so called bulk heterojunction (BHJ) offers a more effective solution for the exciton separation: The active material consist of a penetrating network of the hole and the electron conductor (see figure (2.3)a) which is thick enough to absorb most photons but there are enough interfaces to separate the excitons.

There are basically three different kinds of solar cells which are summarized under the name organic solar cells (OSCs): small molecule solar cells, polymer solar cells and hybrid organic/inorganic solar cells. One of the workhorses in the community of fundamental physical research are the poly(3-hexylthiophene):[6,6]-phenyl-C61-butyric acid methyl ester (P3HT:PCBM) bulk heterojunction (BHJ) solar cells. They can reach efficiencies up to 5% [83]. P3HT is a polymer and acts as the hole conductor with an electron affinity of 2.13 eV [84] and an ionization energy of 4.65 eV [84]. PCBM is a fullerene derivative with an electron affinity of 3.80 eV [84] and an ionization energy of 5.8 eV [84] and conducts the electrons. The molecule structures of both materials are shown in figure 2.4b. The major part of this thesis is about the charge transport in solar cells made from P3HT and PCBM. As the BHJ has a very complex morphology with structure sizes of less than 10 nm [85], it is not

2.3. Organic solar cells

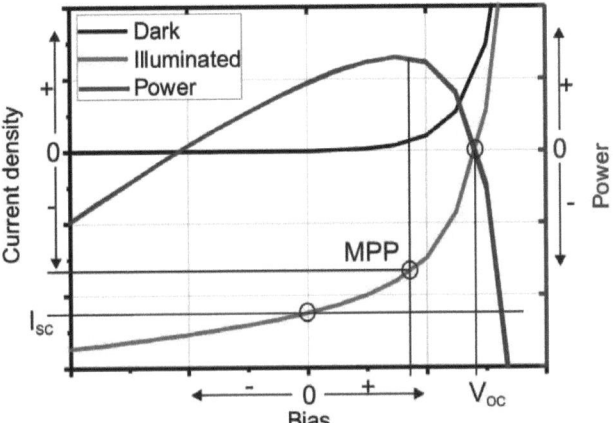

Figure 2.5: IV curve of a solar cell in the dark (black) and under illumination (red). The voltage dependent power of the solar cell is plotted in blue.

possible to resolve morphology effects with SKPM (see chapter 4). Therefore, also bilayer solar cells from P3HT and PCBM were investigated.

Figure 2.5 shows the typical current-voltage (IV) characteristics of a solar cell in the dark (black curve) and under illumination (red curve). The current at zero voltage is called short circuit current (I_{SC}). The voltage at zero current is called open circuit voltage (V_{OC}). The power of the solar cell is given by the negative product of the voltage and the current. In the graph it is plotted in blue. The regime in which the solar cell produces power is called active regime. The state in which the solar cell produces most power is called maximum power point (MPP). One can define a fill factor (FF) by FF=$V_{@MPP} \times I_{@MPP} / V_{OC} \times I_{SC}$.

The efficiency η of a solar cell is given by the produced power at the maximum power point $P_{@MPP}$ divided by the incoming power P_{in}:

$$\eta = \frac{P_{@MPP}}{P_{in}} = \frac{-V_{@MPP} \times I_{@MPP}}{P_{in}} = \frac{-V_{OC} \times I_{SC} \times FF}{P_{in}}$$

3. Preparation of samples

This chapter gives an overview of the sample preparation processes. The first section describes the preparation of TIPS-pentacene OFETs, which have been used for the establishment of SKPM in the InnovationLab. The second section describes the preparation of P3HT:PCBM BHJ solar cells in conventional and inverted device structures. In the third section the fabrication of bilayer solar cells is explained. The major part of the sample preparation has been performed by bachelor, master and diploma students under my mentoring. Therefore, detailed processes and parameters can be found in their theses [40, 86–89].

3.1. TIPS-pentacene OFETs

The 6,13-bis(triisopropylsilylethynyl) pentacene (TIPS-pentacene) [25, 90] OFETs were designed in a bottom gate bottom contact structure. A Si-wafer with 200 nm of thermal oxide on top served as substrate and gate respectively. The contact structure was designed by photolithography and subsequent metal evaporation and lift-off. 60 nm gold was used as contact material with about 10 nm aluminum as adhesive layer. TIPS-pentacene served as the active material of the OFETs and was deposited from solution in different solvents by drop casting or spin coating. The choice of different solvents, the use of surface altering self assembled monolayers (SAMs) and the use of different coating methods lead to entirely different morphologies [86, 91, 92]. The transistor which will be discussed in chapter 6.1 was prepared by drop casting TIPS-pentacene from toluene on an OTS treated substrate.

A detailed description of the preparation is given in the bachelor theses of Lars Müller [86] and Florian Ullrich [87].

3.2. P3HT:PCBM BHJ solar cells

The major part of this work deals with solar cells made from P3HT and PCBM in normal BHJ, inverted BHJ and bilayer structure. In figure 3.1 the different device structures are shown.

3. Preparation of samples

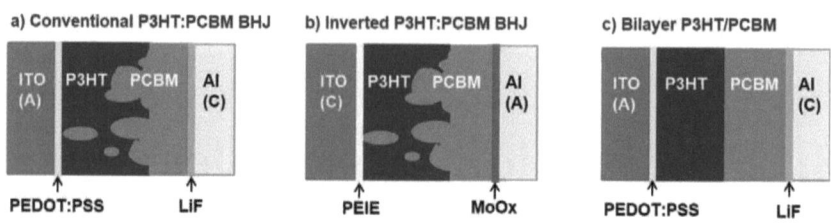

Figure 3.1: Scheme of P3HT:PCBM BHJ solar cells in a) conventional, b) inverted and c) bilayer structure.

The preparation process for our P3HT:PCBM BHJ solar cells in conventional (see figure 3.1a) and inverted (see figure 3.1b) device structure is described in our publication [93]: "BHJ solar cells were fabricated on ITO-coated glass that had been patterned by photolithography to achieve a defined contact structure. The substrate size was chosen to be $5 \times 5\,\text{mm}^2$ as this is the maximum sample size which can be handled within our analytical setup. An approximately 25 nm thick poly(3,4-ethylenedioxythiophene):poly-(styrenesulfonate) (PEDOT:PSS) buffer layer was spin coated on top of the ITO electrode. To prepare active layers of P3HT:PCBM blends, P3HT and PCBM were dissolved in chlorobenzene in a weight ratio of 1:1 with a total concentration of 20 mg/ml for each material. The solution was spin coated onto the PEDOT:PSS coated substrate at a speed of 2000 rpm. As the substrate size was rather small, the resulting layers became inhomogeneous with thicknesses of $2-3\,\mu\text{m}$ at the edge of the substrate and about 200 nm at its center. As top electrode 6 Å LiF and 100 nm Al were thermally evaporated. For the fabrication of the inverted solar cells, polyethylenimine ethoxylated (PEIE) [94–96] was spin coated on top of the patterned ITO substrate. PEIE served as work-function-lowering agent and ITO became the bottom cathode in the inverted OPV architecture. The BHJ was deposited as in the case of the non-inverted solar cells. A 10 nm layer of MoO_3 with 100 nm of Al on top served as the top anode contact. The samples have been prepared under nitrogen atmosphere. They came into contact with ambient air for a few minutes during transport to vacuum. Possible influences of oxygen doping on the potential distribution have been investigated by Morris et al. [33]. Immediately after their preparation, the solar cells were characterized under illuminated (AM 1.5) conditions. They exhibit an open circuit voltage of 5.5 V and a short circuit current of about $8\,\text{mA}/\text{cm}^2$ with a maximum fill factor of 70%. The results for normal and inverted solar cells were similar."

A more detailed description of the preparation process of conventional solar cells is given in the master thesis of Dominik Daume [88] and in the diploma thesis of Michael Scherer [40]. The preparation of inverted solar cells is further described in the bachelor thesis of Julian Heusser [97].

3.3. P3HT:PCBM bilayer solar cells

The preparation of bilayer solar cells (see figure 3.1c) is described in our publication [98]: "The solar cells were prepared on indium-tin oxide (ITO) coated and patterned glass substrates by spin coating of the active materials. First, a poly(3,4-ethylenedioxythiophene):poly-(styrenesulfonate) (PEDOT:PSS) buffer layer was spin coated on top of the ITO electrode. P3HT was solved in chlorobenzene and spin coated with 1000 rpm for 90 s on top of the PEDOT:PSS. As proposed by Ayzner et al. [99] the PCBM was solved in dichloromethane to gain well separated layers. It was spin coated with 4000 rpm for 10 s. From analytical transmission electron microscopy (TEM) measurements [85] on our samples, we certainly know that the P3HT and PCBM form well separated layers. The whole preparation was performed in a glove box under nitrogen atmosphere. 0.6 nm LiF and 100 nm Al were thermally evaporated to form the top contact. Some of our solar cells showed S-shaped IV-curves caused by a temporary iodine contamination of our glove box, which we found out by XPS measurements. We prepared four batches of bilayer solar cells. The first and the third one were prepared in a clean glove box and the second and fourth in a contaminated glove box. In the first and the third batch, all solar cells showed normal IV-characteristics, in the second and fourth batch all solar cells showed S-shaped IV-curves."

An elaborated description of the preparation is shown in the master thesis of Christian Müller [89].

Figure 3.2 shows a transmission electron microscopy (TEM) measurement of the lamella of a P3HT/PCBM solar cell. The lamella has been prepared by FIB milling and subsequent transfer to a TEM grid by use of a micromanipulator [100]. It was prepared from the very same device (MuBiSpin2_3) which will be discussed in chapter 8 after all SKPM measurements had been performed. The left image in figure 3.2 shows an electron-loss measurement at 28 eV where the PCBM appears bright due to its plasmon excitation [85]. The right image shows a zoom into the region of the bilayer interface where the PCBM is colored in red and the P3HT is colored in green. The TEM image shows clearly, that the PCBM and P3HT form a well separated bilayer structure. Lamella preparation as well as TEM measurement and analysis were performed by Diana Nanova [101].

3. Preparation of samples

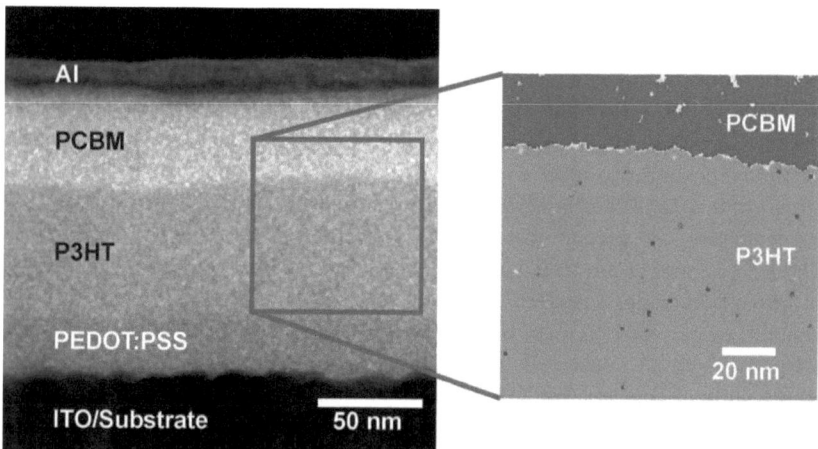

Figure 3.2: TEM measurement of the lamella of a P3HT/PCBM bilayer solar cell. The lamella was prepared by FIB milling and subsequent transfer to a TEM grid by use of a micromanipulator [100].The left image shows an electron-loss measurement at 28 eV where the PCBM appears bright due to its plasmon excitation [85]. The right image shows a zoom into the region of the bilayer interface where the PCBM is colored in red and the P3HT is colored in green. The TEM image shows clearly that the PCBM and P3HT form a well separated bilayer structure. Lamella preparation as well as TEM measurement and analysis were performed by Diana Nanova [101].

4. Measurement methods and experimental setup

The following chapter covers the main measurement methods which have been used in this work. After an introduction into Kelvin probe, scanning probe microscopy, electron microscopy and focused ion beam will be explained. The last section demonstrates the measurement setup that has been used to investigate the potential distribution in solar cells.

4.1. Kelvin probe

Kelvin probe (KP) is a method to determine the contact potential difference (CPD) [102] of a surface relative to the probe of the KP system. The CPD corresponds to the local surface work function (WF) difference between the sample and the probe [103]. The measurement principle is demonstrated in figure 4.1. Sample and probe align to different Fermi levels when there is no electrical contact (figure 4.1a). If they are electrically connected, the Fermi levels align and there is a vacuum level difference which corresponds to the CPD of the two materials (figure 4.1b). In a KP system sample and probe are arranged in a capacitor structure such that different vacuum levels lead to an electric field and a force between sample and probe. This force vanishes if one applies a voltage to the probe which is equal to the CPD (figure4.1c). If a bias voltage is applied to the sample the CPD is altered by this bias voltage.

The detailed control mechanism for conventional KP is explained in [104,105], for SKPM it will be explained in section 4.3. To determine the absolute value of the work function of a material, a reference with very defined work function has to be measured. An adequate reference material forms highly ordered pyrolytic graphite (HOPG). The surface can be refreshed by pulling-off a scotch tape just before the measurement [106].

4. Measurement methods and experimental setup

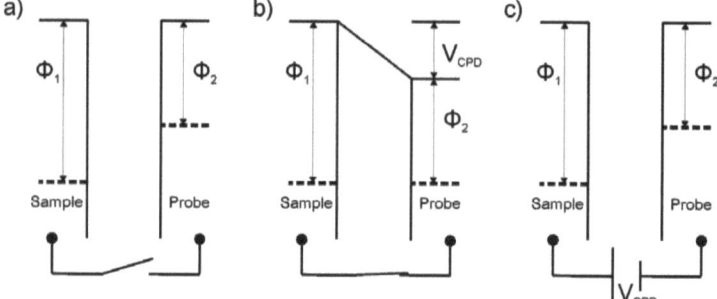

Figure 4.1: The working principle of KP is described: a) The sample with WF Φ_1 and the probe with WF Φ_2 are not in contact. b) Sample and probe are electrically contacted, Fermi level alignmenet leads to a shift of V_{CPD} in the vacuum potential. c) The voltage V_{CPD} is applied between sample and probe such that the vacuum levels align.

4.2. Scanning probe microscopy

The term scanning probe microscopy (SPM) summarizes all methods which perform two dimensional lateral resolved measurements using a probe which scans along the surface of the sample. The most prominent and oldest SPM techniques are scanning tunneling microscopy (STM) and atomic force microscopy (AFM) [107]. AFM offers the possibility to measure the topography of a sample surface with atomic resolution. In AFM the probe consists of a cantilever with typical sizes of $50\,\mu m \times 170\,\mu m$[1]. At its very ending a tip is positioned which extends to the sample. If the probe is approached to the sample at some point an interaction between sample and probe takes place. This interaction leads to a deflection of the cantilever or to a reduction of the oscillation amplitude if the cantilever is oscillating. There are two commonly used types of measuring modes. In contact mode AFM the deflection of the cantilever is kept constant while the probe scans the surface by varying the height position of the sample. In non-contact or tapping mode AFM the reduced oscillation amplitude is kept constant by regulating the sample height during the scan. The non-contact mode is less invasive and therefore preferably used for soft materials like organic semiconductors.

---

[1] We mostly used the cantilever model ATEC-NCPt [108].

4.3. Scanning Kelvin probe microscopy

Scanning Kelvin probe microscopy (SKPM) combines the lateral resolution of SPM with the CPD measurements of KP [109]. There are different possibilities how to realize SKPM. We use a method which is known as one pass amplitude modulated (AM) SKPM. The AFM is operated in the non-contact mode and the SKPM signal is captured simultaneously. An AC voltage with a frequency of a few kHz below the resonance frequency of the cantilever and a DC voltage are applied to the cantilever such that together with the above mentioned CPD a potential between probe and sample is built up:

$$V = V_{DC} + V_{AC}\sin(\omega t) - V_{CPD}. \tag{4.1}$$

As the cantilever and the sample can be considered as a capacitor structure with capacity C, the potential induces a force between them:

$$F = -\frac{1}{2}\frac{dC}{dz}V^2. \tag{4.2}$$

If one inserts the voltage in (4.1) into the term given in (4.2) the total force is conform to:

$$F = -\frac{1}{2}\frac{dC}{dz}\left(V_{DC} + V_{AC}\sin(\omega t) - V_{CPD}\right)^2. \tag{4.3}$$

Solving and sorting the forces with respect to the frequencies such that $F = F_{DC} + F_\omega + F_{2\omega}$ leads to [110]:

$$F_{DC} = -\frac{1}{2}\frac{dC}{dz}\left((V_{DC} - V_{CPD})^2 + \frac{V_{AC}^2}{2}\right), \tag{4.4}$$

$$F_\omega = -\frac{dC}{dz}(V_{DC} - V_{CPD})V_{AC}\sin(\omega t), \tag{4.5}$$

$$F_{2\omega} = \frac{1}{4}\frac{dC}{dz}V_{AC}^2\cos(2\omega t). \tag{4.6}$$

According to equation (4.5) the force which is modulated with frequency

4. Measurement methods and experimental setup

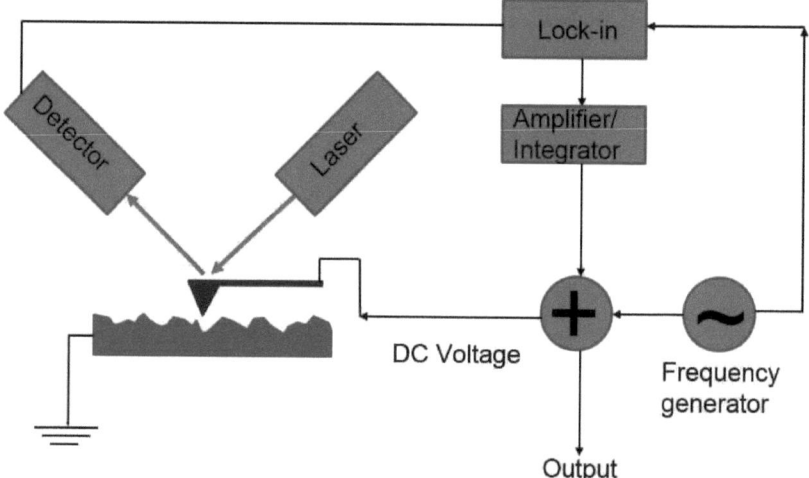

Figure 4.2: Scheme of the SKPM setup: A laser is used to detect the oscillation of the cantilever. The frequency generator provides an AC voltage with frequency ω and with the lock-in amplifier the oscillation of the cantilever with the frequency ω is measured. This oscillation is regulated to zero by adjusting the DC voltage. According to equation 4.5 the DC voltage which regulates the oscillation of the cantilever with frequency ω to zero corresponds to the CPD. Therefore, the DC voltage forms the output. Note that in the scheme the sample is set to ground potential but it can be set on any potential such that it is possible to investigate electronic devices under device operation.

4.3. Scanning Kelvin probe microscopy

ω vanishes if the applied DC voltage is equal to the CPD. Therefore, one regulates the DC voltage such that the cantilever oscillation with frequency ω disappears [111] in order to determine the CPD [109]. In figure 4.2 the scheme of a setup for measuring SKPM is shown. A laser is used to detect the oscillation of the cantilever. The frequency generator provides an AC voltage with frequency ω and with the lock-in amplifier the oscillation of the cantilever with the frequency ω is measured. This oscillation is regulated to zero by adjusting the DC voltage. According to equation 4.5 the DC voltage which regulates the oscillation of the cantilever with frequency ω to zero corresponds to the CPD. Therefore, the DC voltage forms the output. Note that in the scheme the sample is set to ground potential but it can be set on any potential such that it is possible to investigate electronic devices under device operation. Like in KP measurements, the CPD is a relative value and in order to obtain the absolute work function of a sample, one has to use a reference material. In SKPM this standardization is more difficult than in KP because sample mounting is more complex and the tip usually becomes contaminated during the measurement. In this work the relative values and the distribution of the potential is most important. Therefore, the measurements were done without the standardization.

The lateral resolution of AM SKPM typically amounts to 20 nm, the electrical resolution to 5 mV [112]. The lateral resolution is limited to 20 nm as in this SKPM configuration the whole tip and not only its very end interacts with the sample due to long range electrostatic force. The measured image is a convolution of the measured structure and the shape of the tip. Furthermore, the signal depends on the distance between the cantilever and the sample [113]. With high computing time it is possible to some extent to reconstruct the actual surface potential [114].

Moreover, the measured signal is also influenced by the interaction of the whole cantilever with the surface. This interaction actually does not only affect the local resolution but also alters the overall signal, such that the measured CPD difference at different positions of the sample is always smaller than the real CPD difference [115, 116]. Therefore, if a voltage is applied between two contacts, the measured voltage is always smaller [115]. By using coaxial tips and cantilevers, this effect can be avoided [117]. They were not used within this project, as these cantilevers are not yet commercially available. We try to avoid a significant altering of the measured potential in the solar cells by the cantilever, by leaving the top contact on ground potential. However the value of the measured signal is always lower than the applied voltage (compare chapter 7 and chapter 8).

Due to the high spatial resolution, SKPM can be used to map the CPD in

4. Measurement methods and experimental setup

an electronic device. If the device is under operation, the CPD corresponds to a superposition of the local surface work function [103] difference and the effect of the applied voltage [103]. If one subtracts the work function difference from the CPD, the net potential distribution in the device is obtained. The practical approach is explained in chapter 7.

In an electronic device different effects can lead to hindered charge transport (see chapter 2). In figure 4.3a the energetic scheme of the conduction band in an organic device with hindered charge transport due to an injection barrier at the bottom contact and due to crystal grain boundaries or disorder is shown. In a static equivalent circuit such barriers can be displayed as resistances (see figure 4.3b). Note that this eqivalent circuit does not describe the IV curves of the solar cells, it only gives an interpretaion for the potential distribution at a certain voltage. In realistic equivalent circuits of solar cells, one uses at least one diode, a photo current source and some parallel and series resistances [118]. If a bias is applied to such a circuit as shown in figure 4.3 the voltage drops along the resistances (see figure 4.3c). With SKPM the potential distribution under device operation can be measured spatially resolved and the regions of potential drops can be localized. Therefore, one gains information about the major resistances in the device which consequentially gives information about the charge transport barriers. In figure 4.3b the situation is simplified a lot by only considering the series resistances for one certain applied voltage. These resistances are voltage dependent and should only help to understand the conclusions of the measured potential distribution. E.g., a charge transport barrier due to an injection barrier depends strongly on the applied voltage as will be demonstrated in chapter 7.

4.4. Electron microscopy and focused ion beam

In scanning electron microscopy (SEM) an electron beam with acceleration voltage typically in the range of $1 - 30\,\text{kV}$ is scanned along the surface of a sample [119]. The electron beam interacts with the sample and is backscattered, dissolves secondary electrons and stimulates characteristic X-rays. There are different kinds of detectors which detect one of these signals respectively. In our Carl Zeiss AURIGA SEM we use two detectors, the so called in-lens detector which is a secondary electron detector located in the electron lenses and the so called SE2 which is an Everhart-Thornley type detector [120] located a few centimeters besides the electron gun. Our SEM is specified for a maximum

4.4. Electron microscopy and focused ion beam

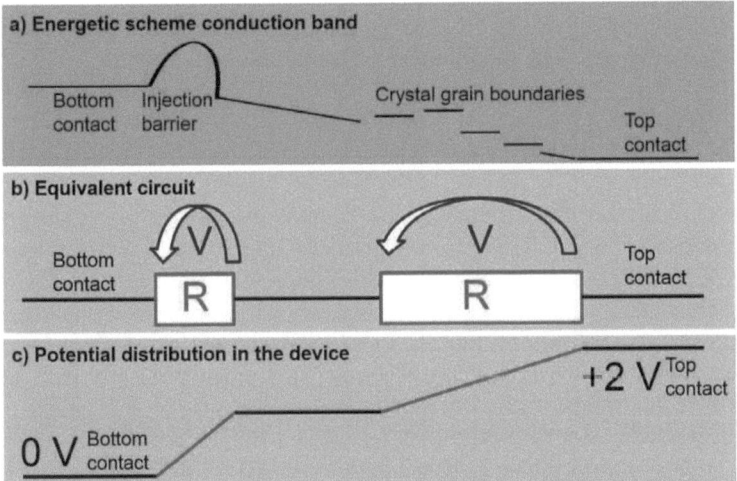

Figure 4.3: Visualization of the measurement concept. a) The energetic scheme of the conduction band in an organic device based on the assumption of an injection barrier at the bottom contact and a hindered charge transport due to crystal grain boundaries or disorder is displayed. b) Equivalent circuit and c) potential distribution in the device. **Note that this equivalent circuit only describes the solar cell in the static case of one certain applied voltage.** The shown resistances are voltage dependent and should only help to understand the conclusions of the measured potential distribution. In realistic equivalent circuits of solar cells, one uses at least one diode, a photo current source and some parallel and series resistances [118].

4. Measurement methods and experimental setup

resolution of 1.2 nm. This maximum resolution is defined as the decline from 80% signal to 20% signal on a material transition with very high contrast. For the certification there was used gold, which has a very high secondary electron emission, and carbon, which has a very low secondary electron emission. That means the resolution in material systems used in this work is far worse and structures usually have to exceed 20 − 50 nm to be resolved.

A focused ion beam (FIB) can be used for imaging as well as for material milling [121]. In our microscope Ga ions are used and they are accelerated with a voltage of 30 kV. The maximum resolution amounts to 7 nm.

4.5. Experimental setup

The main part of the measurements has been performed in a crossbeam SEM/FIB microscope (AURIGA from Carl Zeiss microscopy [122]) with an integrated SPM (BRR from DME [123]). This system allows for preparation of FIB milled cross sections and in-situ measurements with the SPM. Figure 4.4a shows a scheme of the SPM and the SEM and FIB column. The SEM column is placed such that the electron beam is accelerated down vertically and the FIB column is mounted in a 54 degree angle. To readout the movement of the cantilever a red laser is used which is marked by the red line in figure 4.4a. The laser is placed in the laser tube and the beam first hits a movable mirror which directs it via a fixed mirror onto the cantilever. The beam is reflected to another fixed mirror and from there to a movable mirror. This mirror guides the beam to the detector. In figure 4.4b a picture of the SPM in the electron microscope is shown. The SPM unit is fixed to the vacuum chamber door and to take the picture the chamber had to be open. Therefore, in the picture the SPM unit is located in front of the SEM and FIB but with closed door it stands beneath. The upper part shows the components of the standard AURIGA, the SEM column and the FIB column are marked by an arrow. The lower part shows the SPM unit. Note that in this picture there is no sample and no cantilever mounted in the SPM unit. The above mentioned mirror mover and laser tube are marked as well as the scanner plate which moves the sample towards the cantilever and performs the scanning while the cantilever is kept fix. The SPM unit can be moved in all spatial directions and can be tilted towards the FIB up to 80 degree. Rotation of the SPM unit or the sample is not possible. The in plane movement of the SPM is performed by two platters which can move in one dimension each (marked by stage xy in figure 4.4b). The in plane movement of the sample is performed by two forks which hustle the sample into the right position.

4.5. Experimental setup

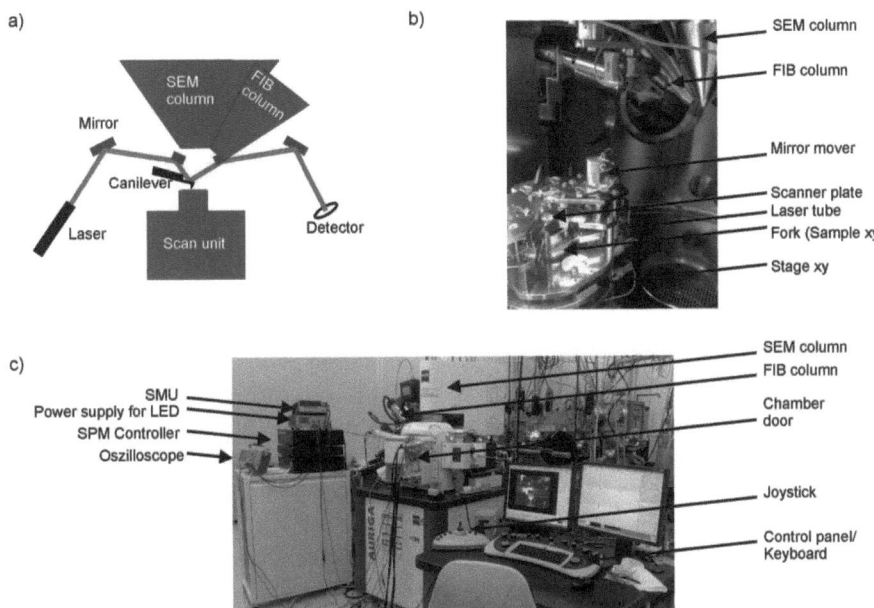

Figure 4.4: a) Scheme of the SPM in the crossbeam microscope. The red line marks the laser path. b) Picture of the SPM in the electron microscope with open vacuum chamber door. c) Picture of the whole measurement setup.

Figure 4.4c) shows the whole measurement setup. On the right, the computer, monitors, control panel and joystick for the SEM, FIB and SPM control are shown. In the middle, the AURIGA microscope with the SPM chamber door is displayed. On the left, the source-measurement unit (SMU) which is used to run the solar cell is shown as well as the power supply for the LED which illuminates the solar cells in the chamber. The oscilloscope is used to monitor the signal coming from the SMU to make sure that it is not noisy. Three black boxes contain the SPM and stage controller.

Figure 4.5a shows a cantilever holder for our SPM. The location of the cantilever is marked by an arrow as well as the pins for the piezo contact and the SKPM signal. If the sample holder is inserted, the pins are connected with electrical feedtroughs such that the driving voltage for the piezo and the SKPM signal can be applied from outside the vacuum system. A resistance of $1\,\mathrm{k\Omega}$ has to be implemented between the pin and the cantilever to avoid electrical breakdowns. Figure 4.5b shows a sample holder for our SPM. During measurement the round plate which is fixed by the two L-shaped wires is lifted

4. Measurement methods and experimental setup

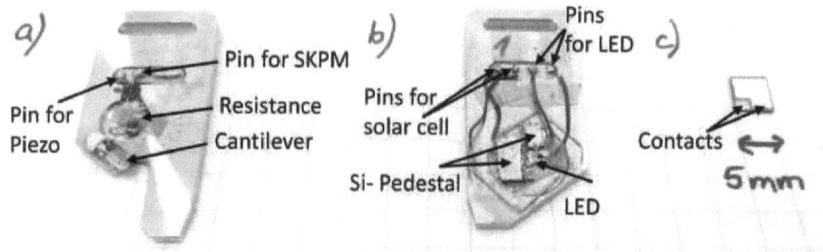

Figure 4.5: a) Cantilever holder, b) sample holder and c) solar cell.

by the scanner and moved with the scanner. On top of this plate an LED is mounted which illuminates the solar cells from the bottom. The solar cell (figure 4.5c) is mounted on the Si-pedestals and fixed with silver glue. Again, the pins are connected to electrical feedthroughs which in this case are used to run the LED and connect the solar cell to the SMU.

Generally, the solar cells were characterized in a solar simulator right after their preparation to obtain the IV curves in the dark and under standard solar AM 1.5 irradiation. Unfortunately, the solar simulator was out of order many times such that IV curves under AM 1.5 illumination are not available for all solar cells. IV curves in the dark and under LED illumination were captured for all cells in the measurement setup. IV curves of the conventional P3HT:PCBM BHJ solar cells were only captured in the measurement setup as at this time we had not yet made an adapter for the solar cell simulator which is restricted for one special cell layout. Additionally, the illumination of the solar cells was performed with a fiber where we incoupled the light of a white light source. The fiber end was positioned at the side of the solar cell, such that there was not very much light coupled to the solar cell. Unfortunately, the original idea of guiding a fiber through the SPM scanner to illuminate from the bottom had to be dismissed as the SPM dimensions were too low and the fiber disturbed the scanning. Nevertheless, there was enough light coupled into the solar cell such that photocurrent and photovoltage were created (compare section 7.2.1).

Once the cantilever holder and sample holder have been mounted to the microscope and the laser path for detection of the cantilever movement has been adjusted, the system gets evacuated. When the vacuum reaches at least $5 \cdot 10^{-5}$ mbar we align the electron and the ion beam at the place where both beams cross (coincidence point). We take IV curves of the solar cells in dark and under illumination before we start milling the devices with the FIB. To

4.5. Experimental setup

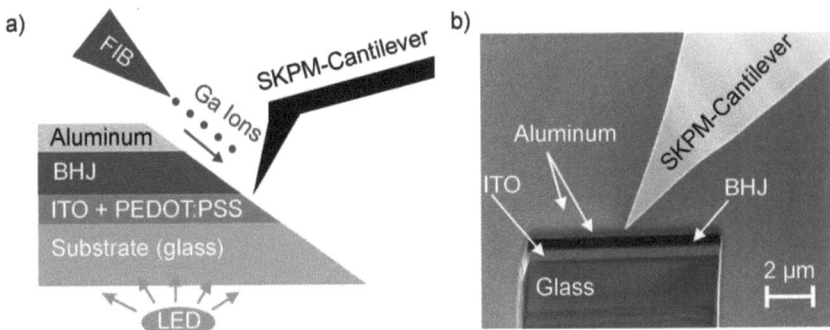

Figure 4.6: a) demonstrates the preparation and measurement method: First, a hole is millled into the solar cell under a 54 degree angle, than the cross section can be measured by SKPM. Additionally illumination of the solar cell can be performed during measurement by the LED. b) shows an SEM image of an FIB milled solar cell which is scanned with a cantilever (From [93]).

expose the cross section, first a coarse milling with 1 nA FIB beam current is performed. In a second step the cross section is polished with a 50 pA FIB beam. In figure 4.6a the preparation and measurement procedure is demonstrated: First, a hole is milled into the solar cell under a 54 degree angle, than the cross section can be measured by SKPM. We used a cantilever with a very sharp tip (cantilever model ATEC-NCPt [108]) to minimize the interaction between the side faces of the cantilever and the FIB milled edge. Illumination of the solar cell can be performed during measurement by the LED. Figure 4.6b shows an SEM image of an FIB milled solar cell which is scanned with a cantilever.

5. Evaluation of the method

The described preparation method is an elegant way to prepare and measure the cross sections *in-situ*. However, it was important to find out to which extend this method alters the properties of the solar cell. First of all, the milling is performed by a Ga FIB which is known to change material properties due to Ga implantation and destruction of molecules [124]. Furthermore, the FIB can dope inorganic semiconductors [125, 126] and we found that doping of organic materials by FIB is also possible [124]. The doping of organic materials by FIB is an unwanted effect for further investigation by SKPM, but doping of organic materials is a widely discussed topic [127–130] as the defined manipulation of electrical properties helps improving devices. Therefore, a systematic doping experiment was performed. In figure 5.1a the setup for this experiment is drafted. TIPS-pentacene OFETs were prepared as described above. The organic layer was deposited by spin coating to provide for uniform layer properties in all devices. The devices were exposed with different Ga doses and the conductivity was measured. As can be seen in figure 5.1b the conductivity can be tuned over six orders of magnitude in a very defined way. By photoelectron spectroscopy and SKPM measurement it was found that the TIPS-pentacene becomes p-doped upon Ga exposure. Further experimental results and discussion can be found in the paper "Doping of TIPS-pentacene via Focused Ion Beam (FIB) exposure" [124].

With the results on doping of TIPS-pentacene the question arose to which extend the FIB milling as preparation method for the cross sections alters the properties of the solar cells. Therefore, the IV characteristics of the solar cells before and after FIB milling had to be compared.

In figure 5.2a the IV curves of a P3HT:PCBM solar cell (C216_3) are shown before (black solid line) and after (red dashed line) FIB milling. It was found that they are not changed by FIB exposure. As the exposed area (ca. $8 \times 8\,\mu m^2$) is significantly smaller than the whole active area of the cell (ca. $2 \times 4\,mm^2$), a local variance of the cell does not need to change the IV curves of the whole device. Therefore, it was important to prepare the cross sections with another method and compare the results. Contrary to most inorganic semiconductors which can be easily cleaved because of their crystallinity, cleaving of organic solar cells leads to very rough surfaces which

5. Evaluation of the method

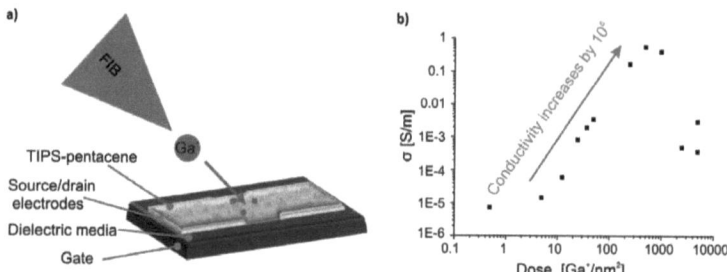

Figure 5.1: a) Scheme of the FIB-doping experiment on TIPS-pentacene OFETs. b) Conductivity increase depending on Ga exposure. From [124].

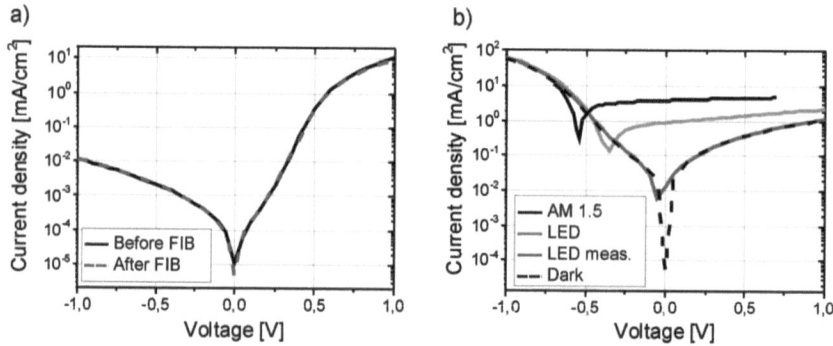

Figure 5.2: a) IV curves of a P3HT:PCBM BHJ solar cell before (black solid line) and after (red dashed line) FIB milling. b) IV curves of an inverted P3HT:PCBM BHJ solar cell (Mu5_1) under different illumination conditions [93].

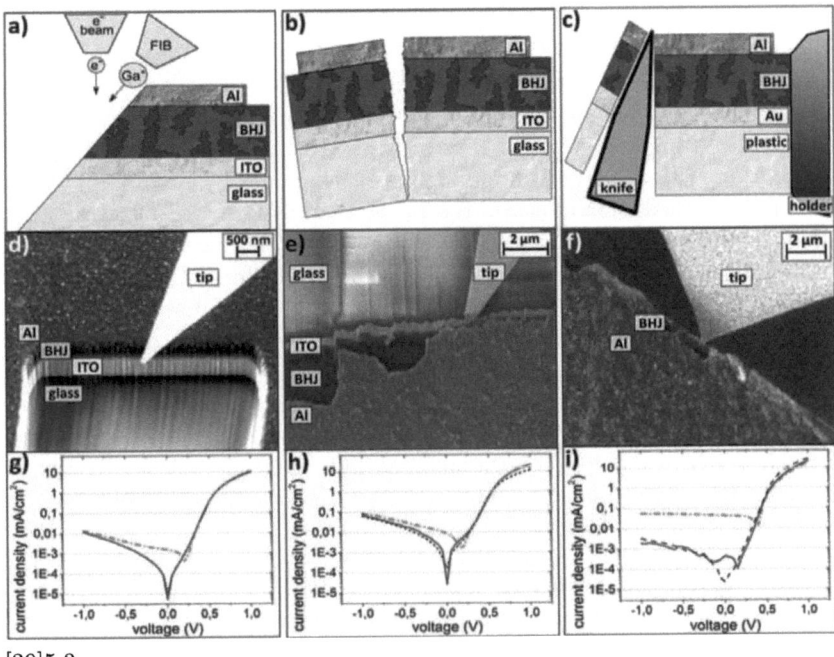

[39]5.3

Figure 5.3: From [39]: "Preparation of OSC cross sections by (a) FIB milling, (b) cleavage and (c) microtome cutting is depicted (view perpendicular to the exposed cross section); SEM images taken during SKPM measurements at (d) FIB milled, (e) cleaved and (f) microtome cut OSC cross sections; Comparison of IV characteristics before and after the sample preparation process by (g) FIB milling, (h) cleaving and (i) microtome cutting. The black (dashed) line indicates the IV characteristic of the untreated solar cell without illumination, the blue (continuous) line indicates the prepared solar cell sample with exposed cross section without illumination and the red (dashed-dot) line indicates the (non-standard) illuminated solar cell sample right before the SKPM measurement".

5. Evaluation of the method

are challenging to measure by SPM. In figure 5.3 the preparation of solar cell cross sections by a) FIB milling, b) cleaving and c) microtome cutting is drafted. Figure 5.3d, e and f show SEM images of the corresponding cross sections. It becomes clear that it is most convenient to measure the FIB milled cross section by AFM as this cross section has the smoothest surface. Note that in this image of the FIB milled cross section, no polish step with low FIB current was performed, therefore the "curtain effect" [131–133] appears very strong. Polished surfaces are smoother (Compare figure 4.6b). Figure 5.3 shows a comparison of IV characteristics before and after the sample preparation process by g) FIB milling, h) cleaving and i) microtome cutting. The black (dashed) line indicates the IV characteristic of the untreated solar cell without illumination, the blue (continuous) line indicates the prepared solar cell sample with exposed cross section without illumination and the red (dashed-dot) line indicates the (non-standard) illuminated solar cell sample right before the SKPM measurement. The IV characteristics were not altered by the preparation of the cross sections.

We managed to capture some SKPM measurements on the cleaved as well as on the microtome cut solar cells. Figure 5.4 shows "surface potential profiles of OSC cross sections under short circuit conditions prepared by (a) FIB milling, (b) cleaving and (c) microtome cutting. Surface potential maps obtained by SKPM are shown in the insets. The displayed profiles were derived from line scans indicated by the arrow. The vertical lines mark the positions of the interfaces between the different layers; d) Surface potential distributions of a cleaved OSC cross section under illumination. The red upper curve shows the short circuit case, the black lower curve shows the case of a floating (disconnected) Al contact (open circuit condition). In the open circuit case the potential of the Al cathode shifted by about $-300\,\mathrm{mV} \approx V_{OC}$. The ITO anode remained on the same potential". [39]. The results show that similar CPD distributions were obtained with all preparation methods. The preparation of cleaved and microtome cut solar cells was mainly done by Michael Scherer who summarized all preparation details and results in his diploma thesis [40]. This cross sections were exposed to ambient air during preparation and mounting. So a contamination by oxygen and water is likely. Therefore, we know at least that FIB preparation does not alter our experiments differently than exposure with ambient air. It will be demonstrated in chapter 7 and 8 that the obtained results for potential distributions of devices under operation lead to coherent conclusions. From transmission electron microscopy (TEM) which was performed on thin FIB prepared lamellas from the very same devices, we know that there is little Ga implantation but the organic materials still show specific behavior [101]. A possible explanation for the FIB having no significant

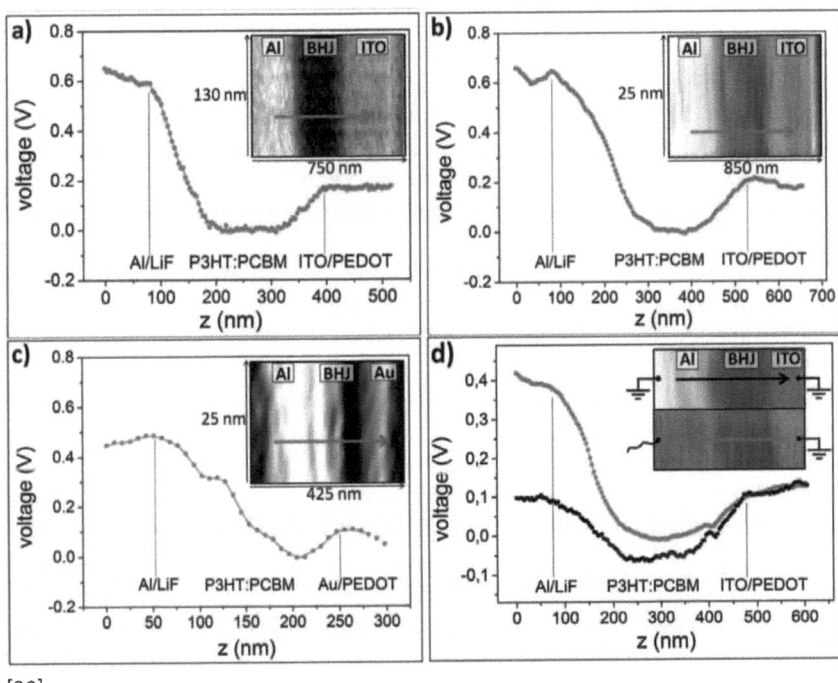

[39]

Figure 5.4: From [39]: "Surface potential profiles of OSC cross sections under short circuit conditions prepared by (a) FIB milling, (b) cleaving and (c) microtome cutting. Surface potential maps obtained by SKPM are shown in the insets. The displayed profiles were derived from line scans indicated by the arrow. The vertical lines mark the positions of the interfaces between the different layers; d) Surface potential distributions of a cleaved OSC cross section under illumination. The red upper curve shows the short circuit case, the black lower curve shows the case of a floating (disconnected) Al contact (open circuit condition). In the open circuit case the potential of the Al cathode shifted by about $-300\,\mathrm{mV} \approx V_{OC}$. The ITO anode remained on the same potential".

5. Evaluation of the method

influence on the measurement results is the fact that the ion beam only streaks the surface of the cross section while most of the organic material lies in the shadow of the metallic top contact. This is different from our experiments on "Doping of TIPS-pentacene via Focused Ion Beam (FIB) exposure" [124] where the organic layer was not covered by any contact.

One can also criticize that the measurements are not performed on rectangular cross sections. Assuming that charge transport occurs vertically to the surface that means that the measurement is not performed directly at the charge transport path. Again we can exclude an effect of this geometric condition as the measurement results are similar in the cleaved and microtome cut devices where the cross sections were approximately rectangular to the surface [39].

Overall, it cannot be excluded that the cross section preparation affects the properties of the surface of the cross section but there are many hints which argue against a significant influence on our measurement under device operation.

The measurement procedure has another deficit: we are able to illuminate the solar cells during measurement with an LED but due to the small dimensions, it is not possible to illuminate with standard AM 1.5 light. During measurement it was not possible to run the LED on full power as heating of the solar cells leads to drifting (see section A.1.3). In figure 5.2b IV curves of an inverted P3HT:PCBM BHJ solar cell (Mu5_1) under different illumination conditions are shown. The dark IV curve is displayed by the black dashed line, the IV curve under AM 1.5 illumination by the solid black line and the green line shows the IV curve under LED illumination. The red line displays the IV curve under reduced LED power. The figure shows that not only the current but also the open circuit voltage depends on the illumination which has to be taken into account during analysis of measurement results.

6. SKPM on OFETs and Si solar cells

As research at InnovationLab just started a few month before the start of this thesis, the measurement equipment first had to be installed and launched. Furthermore, SKPM is a well established method but the correct choice of measurement parameters and the interpretation of results requires some experience. Therefore, we decided to start with measurements on well investigated devices and confirm literature results. In this chapter the first experimental results are described and discussed. Measurements will be shown which have also been performed by other scientific groups to proof that the establishment of SKPM in our laboratory was successful. The first section will be concerned with the investigation of a TIPS-pentacene OFET. The potential profiles were measured with a standard SPM system in ambient air. In the second section the analysis of a standard Si solar cell will be presented. This study was performed with the combined cross beam/SPM system.

6.1. TIPS-pentacene OFET

OFETs have been investigated with SKPM by many different groups [30,34,35] as the charge transport occurs parallel to the surface and therefore is easily accessible by SKPM. As first test devices we chose OFETs with TIPS-pentacene as an active layer (see chapter 3). The investigation of these devices was performed during spring-time in 2011 before the delivery of our combined SEM/FIB and SPM microscope which is described in chapter 4. Therefore, a leased SPM from DME [134] which is capable of measuring SPM at ambient air was used.

Figure 6.1 shows topography images of TIPS-pentacene OFET channels. Figure 6.1a is taken from the PhD thesis of Stephen Bain, University of Southampton, who prepared the active layer by zone casting [135] of TIPS-pentacene from solution in mesitylene [136]. Figure 6.1b shows one of our devices which was prepared by drop casting TIPS-pentacene from toluene on a OTS treated substrate. The obtained morphologies are similar with elongated

6. SKPM on OFETs and Si solar cells

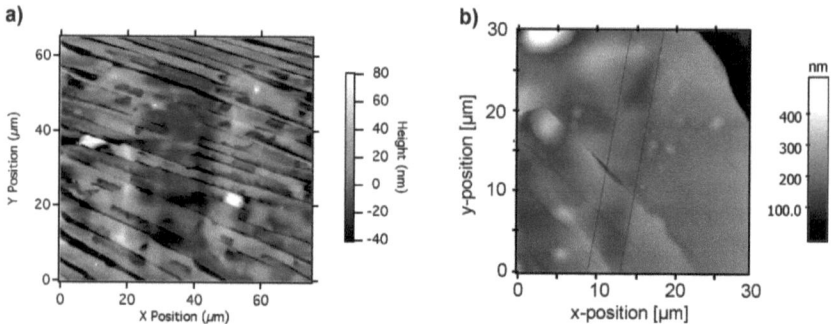

Figure 6.1: Topography image of a TIPS-pentacene OFET a) from [136] and b) measured within this project.

crystals covering the channel in an arrangement nearly parallel to the charge transport. Note that the images have different scales such that the crystals in figure 6.1b appear broader although they have approximately the same size as the crystals in figure 6.1a. The location of the channel can be estimated from light microscope measurements and is marked in figure 6.1b by the black lines.

Figure 6.2 shows the corresponding surface potential measuremets. In Figure 6.2a) a source-drain voltage (V_{SD}) of 5 V and a gate voltage (V_G) of -40 V were applied [136]. In Figure 6.2b) the gate voltage amounted to zero and the source-drain voltage to 5 V. In both images one can see that the voltage drops from the electrode with the higher potential to the electrode with the lower potential inside of the channel. The potential distribution is influenced by the crystal grain boundaries in both measurements.

The good agreement between the results shows that the establishment of SKPM has been successful. The above shown measurement is only one example for the potential distribution within OFETs. Different morphology or contact properties can change the potential distribution significantly.

Further results and discussion on the morphology and the potential distribution of the TIPS-pentacene OFETs which have been investigated in the framework of this project can be found in the Bachelor theses of Lars Müller [86] and Florian Ullrich [87].

6.2. Cleaved Silicon solar cells

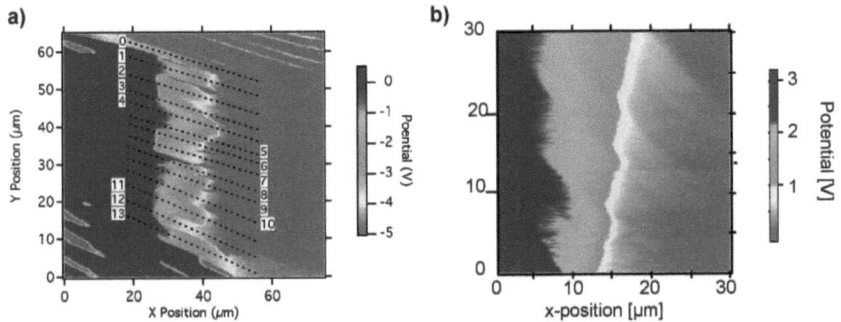

Figure 6.2: SKPM image of a TIPS-pentacene OFET with a) $V_{SD} = 5\,\text{V}$ and $V_G = -40\,\text{V}$ [136] and b) $V_{SD} = 5\,\text{V}$ and $V_G = 0\,\text{V}$.

6.2. Cleaved Silicon solar cells

In inorganic semiconductor physics the investigation of device cross sections by scanning (Kelvin) probe microscopy is a well established technology and has been performed by many groups on LEDs [137] as well as on solar cells [138–140] and other inorganic structures [141]. As inorganic solar cells usually consist of crystalline material, the cross section can be easily accessed by cleaving of the device. Therefore, standard silicon solar cells were purchased at a shop for electronic equipment[1]. Small pieces with a height of about 2 mm were cleaved from the solar cells and mounted on the sample holder. The cleaved cross section formed the face. In SEM images [89] the pyramids which result from anisotropic KOH etching [142, 143] were visible. So it can be concluded, that the solar cells were monocrystalline. Due to this very high pyramids compared to the size of cantilever and to the maximal scan size, it was not possible to measure SKPM on FIB milled cross sections. But measurements on the cleaved cross sections worked well. In figure 6.3 the CPDs of two Si solar cells without applied bias voltage are shown. Figure 6.3a is taken from the paper of A. Breymesser et al [144]. In this figure the surface potential (CPD) is plotted as a black line which refers to the right y-axis. From this potential distribution the electric field strength is deduced and plotted in grey (refers to the left y-axis). Figure 6.3b was measured in our system. The shape of both black curves appear very similar with a higher CPD signal in the n-doped region than in the p-doped region. This is a reasonable result as the work

---
[1]Conrad, Sol Expert Solarzellenbruch Nennspannung 0,45 V Nennstrom 3 A, order number: 110154 - 62

6. SKPM on OFETs and Si solar cells

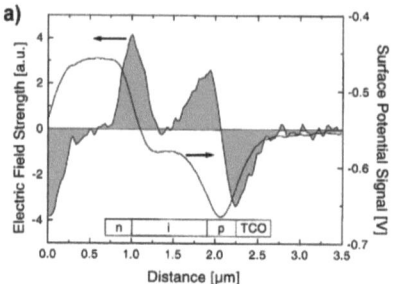 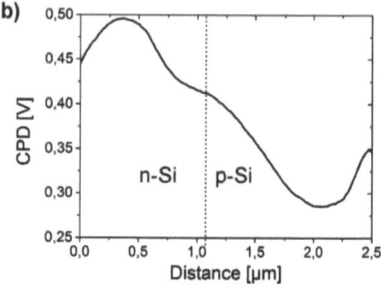

Figure 6.3: Surface potential (CPD) of a a) microcrystalline Si solar cell [144] and b) monocrystalline Si solar cell. In a) the electric field strength is deduced from the surface potential and plotted in grey.

function of n-doped Si is lower than that of p-doped Si. The absolute values of the CPD do not match as they are dependent on the tip material. The maximum difference of the measured CPD in the p-doped and the n-doped region amounts to 210 mV in both cases.

In figure 6.4 potential profiles within two Si solar cells with applied bias voltages are shown.

The signal was captured by measuring the CPD on the cross section under application of different bias voltages. Afterwards, the signal at 0 V was subtracted from the signals at different voltages (this procedure is explained in detail in chapter 7). In figure 6.4a the anode was kept on constant potential while the potential on the cathode was altered. The applied bias voltage drops along the depletion zone with flat slope in the n-doped region and steeper slope in the p-doped region. In figure 6.4b the cathode was kept on constant potential and different bias voltages were applied to the anode. So at a first glance the profiles appear different but again, the voltage drops along the transition between the n-doped and p-doped region with a flat slope in the n-doped region and a steeper slope in the p-doped region. The depletion region in the p-doped zone, which means the distance between the p-n junction and the point of constant potential, is different in both measurements. In figure 6.4a the depletion zone extends from the p-n junction to the point where the profiles run together which amounts to appr. 1.5 μm. In figure 6.4b it extends from the p-n junction to the point where the profiles run parallel which amounts to appr. 0.7 μm. This difference might result from the fact that in figure 6.4a the bias voltage is applied in reverse direction and therefore a broader depletion zone is expected. Differences in the depletion zone length

6.2. Cleaved Silicon solar cells

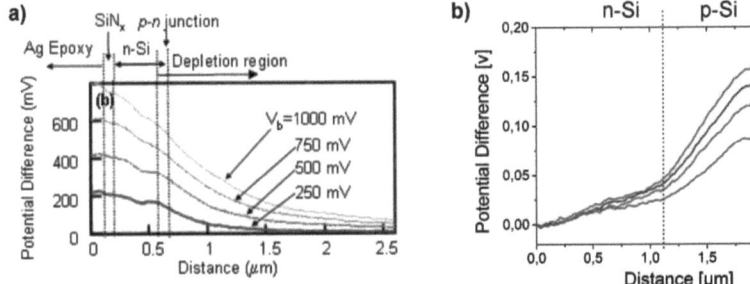

Figure 6.4: Potential difference in Si solar cells under different applied bias voltages. a) is taken from [145] and the anode was kept on constant potential while the potential on the cathode was altered. b) was captured with our microscope. The cathode was kept on constant potential and different bias voltages were applied to the anode.

can also occur from different doping or from different measurement geometries.

The above shown results confirm that with our experimental equipment and procedure it is possible to reproduce potential measurement within Si solar cells. Further measurements and interpretations for the Si solar cells are discussed in the Master's thesis of Christian Müller [89].

7. Potential distribution within P3HT/PCBM BHJ and bilayer solar cells

In this chapter the measurements on different P3HT/PCBM solar cells are shown and discussed. In chronological order, the BHJ solar cells were investigated before the bilayer solar cells as only after the measurements of the BHJ it became clear that a closed description is only possible by removing the complex morphology. However, to simplify interpretation, the results on the bilayer cells are shown in the first section. The second section is about conventional BHJ solar cells and the third section is about inverted BHJ solar cells. The nomenclature of the different samples is intensionly left as it was originally to allow for traceability on the basis of the laboratory books.

7.1. P3HT/PCBM bilayer solar cells

Using the example of P3HT/PCBM bilayer solar cells the procedure of data acquisition is explained in detail.

7.1.1. IV curves

During the investigation of the bilayer solar cells, a functional solar simulator was not available such that only IV curves in the dark and under LED illumination were taken. Figure 7.1 shows IV curves of two bilayer P3HT/PCBM solar cells (MuBiSpin3_1 and MuBiSpin3_2) in a) linear and b) logarithmic scale. The IV curves show that both solar cells have similar characteristics with open circuit voltage of about 0.5 V and short circuit current of about 1.5 mA/cm² with fill factors of about 52% under LED illumination. For MuBiSpin3_2 an open circuit voltage of 0.51 V and a short circuit current of 4.75 mA/cm² were taken under ambient, afternoon sun exposure [1]. MuBiSpin3_1 was already

---
[1] Taken on July 19, 2013 at 3pm

7. Potential distribution within P3HT/PCBM BHJ and bilayer solar cells

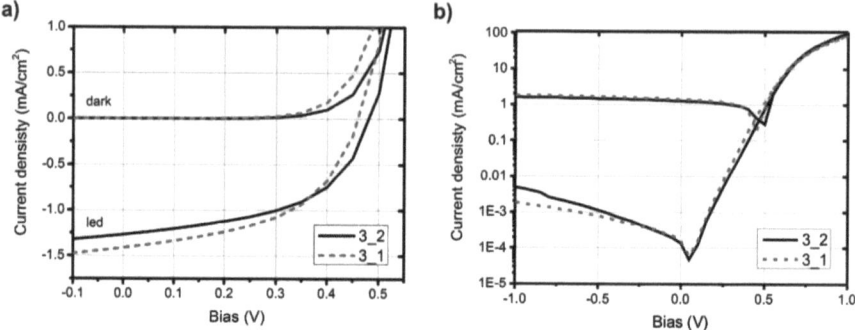

Figure 7.1: IV curves of two bilayer P3HT/PCBM solar cells (MuBiSpin3_1 and MuBiSpin3_2) in the dark and under illumination in a) linear and b) logarithmic scale.

degraded because of too many measurements in ambient air, but another solar cell from the same batch (MuBiSpin3_3) was measured as well and showed an open circuit voltage of 0.50 V and a short circuit current of 5.50 mA/cm². In our experiments it was most important that the cells show reasonable open circuit voltages and short circuit currents. It was not attempted to fabricate cells with highest efficiencies but the focus was on comparing different properties and evaluating the method.

7.1.2. Analysis of the cross section

The cross sections were prepared by FIB milling as explained above. After milling the cross section, the SPM cantilever was positioned at the region of interest by using the SEM. To minimize beam damage or carbon deposition on the active layer, the SEM was used with low beam current and short exposure time. Therefore, high resolving SEM images were only captured after the SPM characterization of the surface. Figure 7.2a shows the SEM image of the cross section of a bilayer solar cell. The different materials are marked in the image. ITO and Al appear as bright layers in the image. The organic layer (marked as „O") always appears very dark. A differentiation of P3HT and PCBM is not possible as both mainly consist of carbon which, if not highly doped, emits only few secondary electrons. Note that the PEDOT:PSS layer is not marked. It has a thickness of only a few nanometers and a similar contrast as ITO. So with the secondary electron detector it cannot be resolved. The LiF layer has only a thickness of 0.7 nm which is smaller than the maximum resolution of

7.1. P3HT/PCBM bilayer solar cells

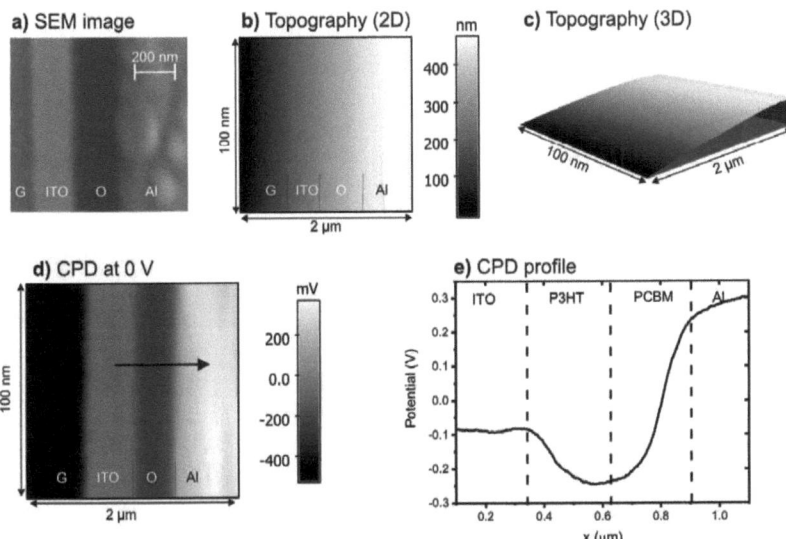

Figure 7.2: a) SEM image, b) topography in 2D representation and c) topography in 3D representation of the cross section of MuBiSpin3_2. The scale bar refers to both topography images. d) SKPM measurement (spatially resolved CPD) of MuBiSpin3_2 without applied bias voltage. The arrow marks the position where the profile shown in e) was taken. In the profile in e) the position of the ITO, the P3HT/PCBM bilayer and the Al is marked.

the SEM (1.2 nm). In figure 7.2b the AFM image of the cross section is shown in a 2D representation and the different layers are marked. In figure 7.2c the topography is presented in 3D. The scale bar between figure 7.2b and c refers to both images. The topography images show that the FIB milling yields a smooth and continuously declining cross section.

In figure 7.2d the CPD image without applied bias voltage is shown. The different layers can clearly be distinguished by their different CPDs. Note, that the PEDOT:PSS layer cannot be resolved as its thickness is below SKPM resolution of 20 nm (compare section 3 and 4.3). Same is valid for the LiF. Therefore, in all further figures and descriptions the position of the ITO and the Al will be indicated and the PEDOT:PSS will be treated as part of the ITO contact. However, the presence of PEDOT:PSS (or PEIE in the case of inverted cells) has to be kept in mind as these materials significantly influence the energy level alignment and the device properties. Figure 7.2e shows a profile of the SKPM signal along the solar cell's cross section. The position

7. Potential distribution within P3HT/PCBM BHJ and bilayer solar cells

where the profile was taken is marked by an arrow in figure 7.2d. In the graph, the position of the ITO, of the Al and the estimated position of P3HT and PCBM are marked. The CPD of ITO is measured to be around 0.4 V smaller than the CPD of Al. Therefore the measured relative work function of ITO is higher than the work function of Al. Regarding all measurements on cross sections of different P3HT/PCBM solar cells, the difference of the CPD of Al and ITO was in the range between 0.4 V and 0.6 V. As explained in section 4.3 the measured signal is always influenced by the interaction of the cantilever with the sample and so the measured potential difference is always smaller than the real potential difference. This effect is more or less pronounced depending on the geometry and the measurement condictions. Therefore, the CPD difference is not always the same and it can be assumed that the real difference is slightly higher than the measured difference. In literature one finds different values for the work function of ITO and Al, highly depending on the processing parameters and surface treatment. The literature values of the work function of ITO vary between 4.4 eV [74, 146] and 5.5 eV [147]. For Al there can be found literature values around 4.2 eV [46]. Anyway, it can be stated that the work function of ITO is higher than the work function of Al which also complies all measurements performed within this project. Furthermore, in all measurements the P3HT and PCBM had a lower CPD than the ITO which means a higher work function. Most work function measurements of P3HT and PCBM which can be found in literature were performed by photoelectron spectroscopy. For P3HT one finds 3.9 eV [74] and for PCBM 4.2 eV [74]. The ionization energy of P3HT amounts to 4.65 eV [84] with a band gap of 2.52 eV [84] and the ionization energy of PCBM amounts to 5.80 eV [84] with a band gap of 3.0 eV [84]. Therefore, a higher work function than reported by Xu et al. [74] is possible. With conventional KP Wu et al. found 4.4 eV as the work function in a P3HT:PCBM BHJ [148]. With the KP system in our laboratory, Tobias Jenne measured a PCBM work function of 5.2 eV and a P3HT work function of 4.6 eV [149]. As reference a highly ordered pyrolytic graphite (HOPG) sample was used. The surface was refreshed by pulling-off a scotch tape a few minutes before the measurement [149]. In the SKPM measurements on cross sections of all P3HT/PCBM solar cells (bilayer and BHJ) the relative work function of the organic layer was always about 0.2 V higher than the relative work function of the ITO. This value lies within the range of the above mentioned values. Altogether, the measured CPD differences are lower than they are in reality due to interaction of the whole cantilever with the sample (compare section 4.3). The P3HT and PCBM cannot be distinguished in the bilayer devices as well. As the layers are very small the influence of the depletion zone cannot be neglected. In thicker bilayer

7.1. P3HT/PCBM bilayer solar cells

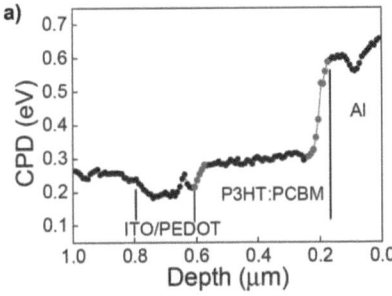 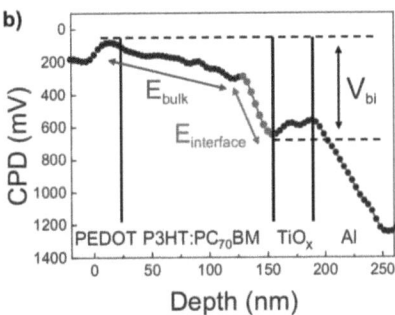

[150, 151]

Figure 7.3: a) From [150]: "The mean-field potential line-profile of the cross-sectional image". b) From [151]: "The linear potential profile of the device in a short circuit condition".

devices a differentiation of the work functions should be possible as long as the work function difference exceeds about 50 − 100 meV.

As far as published in literature, only one other scientific group has measured SKPM on cross sections of organic solar cells. Jongjin Lee, Jaemin Kong et al. cleaved and investigated a standard P3HT:PCBM solar cell under nitrogen atmosphere [150]. In figure 7.3a the CPD profile which they measured is demonstrated. Similar to our results, the interface between the Al contact and the BHJ can be clearly resolved by a CPD drop and the work function of the BHJ is higher than the work function of the Al. They obtained similar relative work function values for the Al and ITO but the organic layer had a slightly lower work function than the ITO. However, not all experimental details were described in their publications. Especially the surface roughness remained unclear. In rare events, our measurements showed similar behavior but we could always identify a too rough surface, or a loose contact or a contaminated cantilever, as the main reason for the observed differences. Furthermore, Jongjin Lee, Jaemin Kong et al. did not make any statement about the reproducibility of their results. From our experiences, the investigation of cleaved cross sections is very complex, time-consuming and less reproducible. In a further publication of this group, the cross section of a P3HT:[6,6]-phenyl-C71-butyric methyl ester ($PC_{70}BM$) BHJ solar cell with a TiO_x/Al top contact was investigated. The obtained CPD profile is plotted in figure 7.3b. Note that the y-axis is turned upside down. This time, the organic layer showed a higher work function than the ITO. $PC_{70}BM$ has an ionization potential of 6.1 eV [152]

45

7. Potential distribution within P3HT/PCBM BHJ and bilayer solar cells

which is higher than the ionization potential of $PC_{60}BM$ (5.8 eV [84]). But, a variation of 0.3 eV in the ionization energy of one component can hardly explain a shift in the work function of the BHJ by 0.3 eV. Therefore, it is more likely that the accuracy of the measurement is limited by the surface roughness. Again, the reproducibility is not commented. In principle their measurements show same tendency as our measurements but they also show that SKPM on a cleaved surface is much harder to perform and to interpretate.

In the following investigations of other P3HT/PCBM solar cells SEM, AFM and SKPM images without applied bias voltages will not be shown as they were all similar to the measurements discussed in this section.

7.1.3. SKPM data with applied bias voltage in the dark

Normally, the procedure for measuring the potential distribution under applied bias voltage was carried out the following way: For each bias voltage a scan field of $1-2\,\mu m \times 100\,nm$ was measured. Figure 7.4 shows the SKPM measurements of the cross section of a P3HT/PCBM bilayer solar cell (MuBiSpin3_2) under different applied bias voltages: a) -1 V, b) -0.5 V, c) 0 V, d) 0.5 V, e) 1.0 V, f) 0.4 V, g) 0.3 V, h) 0.2 V, i) 0.1 V, j) 0 V, k) -0.1 V, l) -0.2 V, m) -0.3 V, n) -0.4 V. Between -0.5 V and 0.5 V fine steps were used to get very detailed results in the region of the open circuit voltage. The scan field size amounts to $2\,\mu m \times 100\,nm$ and the scale bar shows the value of the CPD in mV. Each image took about one hour to capture, as one had to measure with a velocity of only $0.25\,\mu m/s$ to obtain high SKPM resolution in the vacuum. The measurements were carried out in the same order as they appear in the figure. Usually the components in the microscope drifted such that it was often necessary to adjust the measurement and to wait until the next day until everything was settled. In most cases the first measurements still drifted strongly which can be seen in figure 7.4a. Therefore, during the next measurement (figure 7.4b) the position had to be adjusted. As the system was still not completely settled, in figure 7.4d the position had to be adjusted again. In figure 7.4e it can be seen that the drift occurred not only parallel to the scan direction but also orthogonal such that the cantilever hit the frame of the FIB milled hole. Therefore, between scan e and f the position vertical to the scan direction was also adjusted. The shown series of images represents a typical result out of several similar ones. There were measurements with no drift and the best possible resolution but there were also measurements with even more drift such that sometimes it was not possible to extract data.

From these scan fields profiles were extracted. This was done by hand as the drift did not allow for an automatized procedure. Figure 7.5 shows the

7.1. P3HT/PCBM bilayer solar cells

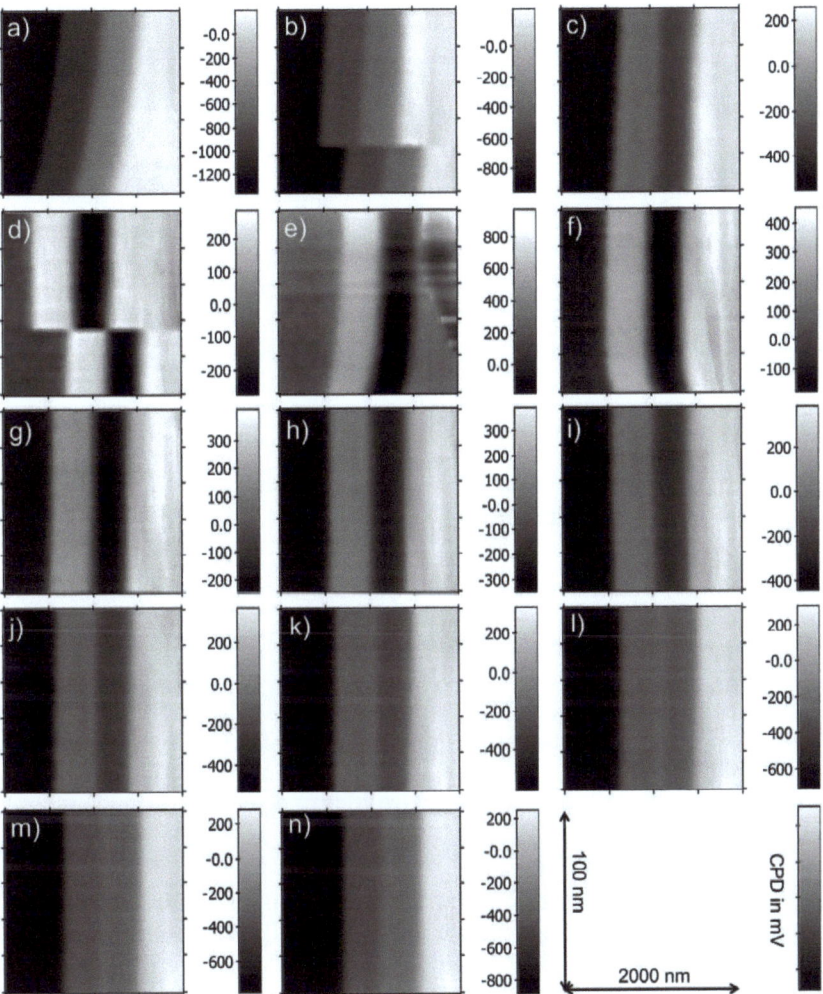

Figure 7.4: SKPM measurements of the cross section of a P3HT/PCBM bilayer solar cell (MuBiSpin3_2) under different applied bias voltages in the dark: a) -1.0 V, b) -0.5 V, c) 0 V, d) 0.5 V, e) 1.0 V, f) 0.4 V, g) 0.3 V, h) 0.2 V, i) 0.1 V, j) 0 V, k) -0.1 V, l) -0.2 V, m) -0.3 V, n) -0.4 V. The scan field size amounts to $2\,\mu m \times 100\,nm$ and the scale bar shows the value of the CPD in mV. The Al contact was set on ground potential and one can observe the shift in surface potential of the ITO contact.

7. Potential distribution within P3HT/PCBM BHJ and bilayer solar cells

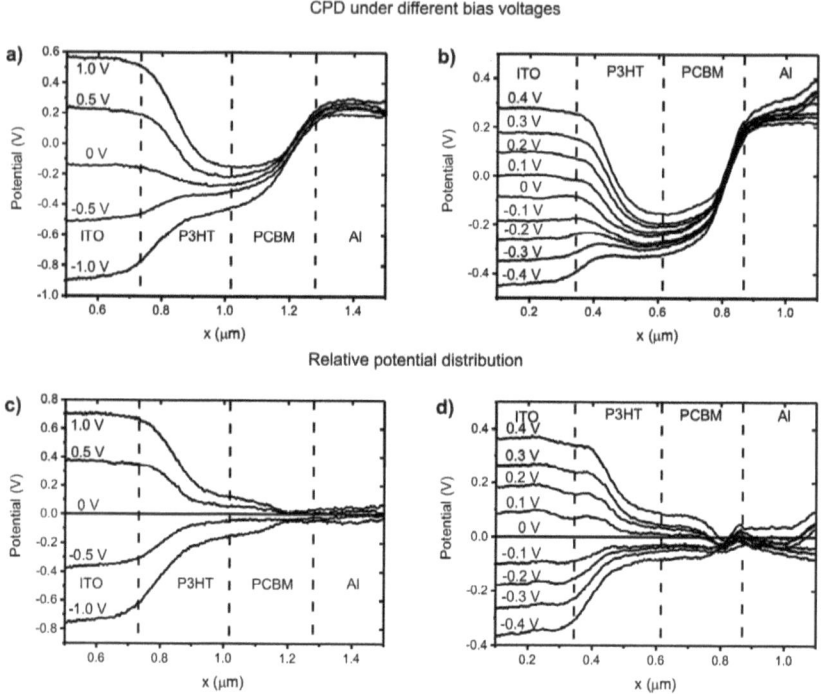

Figure 7.5: CPD under different bias voltages in a bilayer P3HT/PCBM solar cell (MuBiSpin3_2) in the dark for a) $-1\,\text{V} \to 1\,\text{V}$ in $0.5\,\text{V}$ steps and b) $-0.4\,\text{V} \to 0.4\,\text{V}$ in $0.1\,\text{V}$ steps. c) and d) show the respective relative potential distributions.

respective profiles. In figure 7.5a the CPD is shown for $-1\,\text{V} \to 1\,\text{V}$ in $0.5\,\text{V}$ steps and in figure 7.5b it is shown for $-0.4\,\text{V} \to 0.4\,\text{V}$ in $0.1\,\text{V}$ steps. The position of the ITO, of the organic layers and of the Al are marked. As already mentioned, the measured CPD represents the absolute local surface potential which is a super position of the work function and the applied voltage. Therefore, to gain the net voltage distribution in the device, one has to subtract the part of the work function. If there is no bias voltage applied, only the work function is measured. So, the measurement at $0\,\text{V}$ bias voltage corresponds to the work function distribution.

In figure 7.5c and d the $0\,\text{V}$ line is subtracted, respectively. The potential is constant on the ITO and the Al, due to their high conductivity no voltage drop occurs. Between the two contacts the voltage drops off steadily. The

7.1. P3HT/PCBM bilayer solar cells

major part of the voltage drops at the contact between ITO and P3HT. A minor part drops across the bilayer. There is no drop at the interface between PCBM and Al. That means that the major barriers for charge transport in this device are formed by the bottom contact and the organic material itself whereas the LiF/Al contact does not disturb the charge transport measurably.

7.1.4. SKPM data with applied bias voltage under illumination

Data acquisition with applied bias voltage was also performed under LED illumination. In figure 7.6 the data of the bilayer solar cell MuBiSpin3_2 is shown and in figure 7.7 it is shown for the bilayer solar cell MuBiSpin3_1. Under the applied illumination, the solar cells exhibited an open circuit voltage of ca. $0.4\,V$. In figure 7.6a the CPD is shown for $-1\,V \to 1\,V$ in $0.5\,V$ steps and in figure 7.6b it is shown for $-0.4\,V \to 0.4\,V$ in $0.1\,V$ steps. Again, the $0\,V$ line was subtracted and the respective net potential distributions are shown in figure 7.6c and d.

As a comparison the potential profiles of a second bilayer solar cell MuBiSpin3_1 under illumination are presented. The corresponding CPD profiles are shown in figure 7.7. Again, in figure 7.7a the CPD is shown for $-1\,V \to 1\,V$ in $0.5\,V$ steps and in figure 7.7b for $-0.4\,V \to 0.4\,V$ in $0.1\,V$ steps. Figure 7.7c and d show the respective net potential distributions.

In both solar cells the main part of the applied voltage drops at the interface between ITO and P3HT, a minor part drops across the organic layer and no voltage drops at the interface between PCBM and Al. The profiles are similar to the profiles measured in the dark which means that the contact properties and the conductivity of the organic material do not change upon illumination or do change to the same amount.

7.1.5. SKPM data in open circuit condition

Certainly one of the most important issues to investigate with our method is the distribution of the open circuit voltage. Therefore, experiments were conducted in which the CPD was measured in the dark and under illumination in short circuit and open circuit condition. Figure 7.8a shows such an experiment. A scan field was captured and during the scan the conditions were changed. The scan direction was from the bottom to the top in horizontal lines. The glass substrate, the ITO, the organic layer („O") and the Al contact are marked in the scan field. The scan started in region A: both contacts were grounded

7. Potential distribution within P3HT/PCBM BHJ and bilayer solar cells

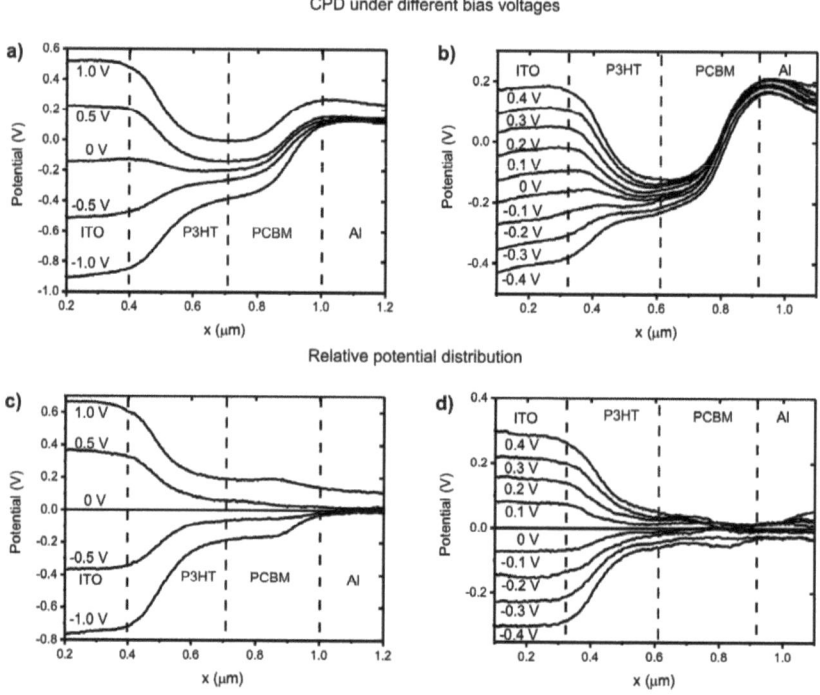

Figure 7.6: CPD under different bias voltages and under illumination in a bilayer P3HT/PCBM solar cell (MuBiSpin3_2) for a) −1 V → 1 V in 0.5 V steps and b) -0.4 V → 0.4 V in 0.1 V steps. c) and d) show the respective relative potential distributions.

7.1. P3HT/PCBM bilayer solar cells

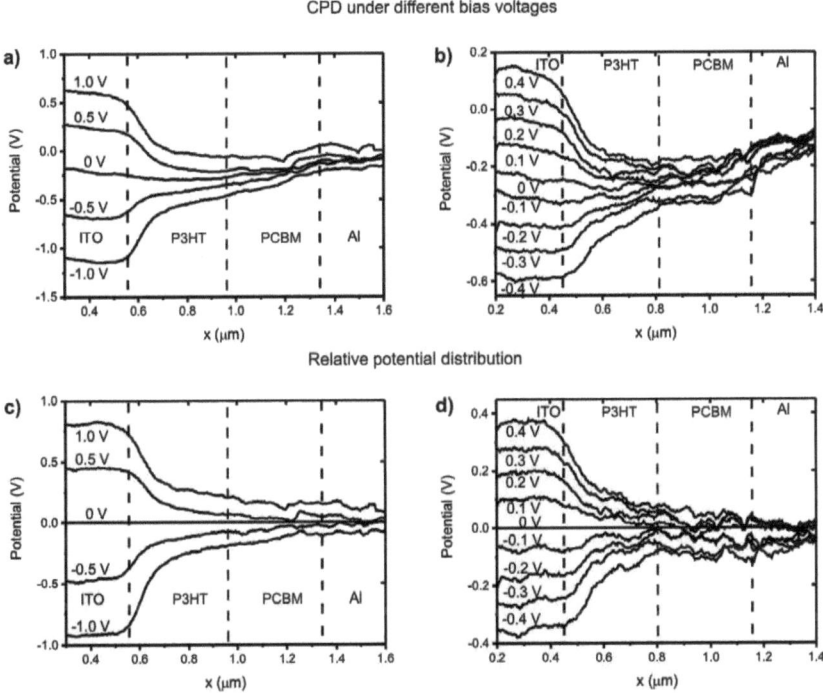

Figure 7.7: CPD under different bias voltages and under illumination in a second bilayer P3HT/PCBM solar cell (MuBiSpin3_1) for a) −1 V → 1 V in 0.5 V steps and b) -0.4 V → 0.4 V in 0.1 V steps. c) and d) show the respective relative potential distributions.

7. Potential distribution within P3HT/PCBM BHJ and bilayer solar cells

(short circuit condition) and the illumination was turned off. In region B the sample illumination was turned on but the sample was kept in short circuit condition. The measured signal did not show any changes but the scan started drifting to the left. This happened due to heating of the sample caused by the illumination and the photocurrent in the solar cell. The heating of the sample causes a slight swelling which is strong enough to disturb the accuracy of the SKPM measurements. To avoid the drift due to heating during such time consuming measurement shown in figure 7.6 and 7.7 there are two possibilities: One could wait until the swelling stops and an equilibrium is reached or one can use less incident light power. In this measurement we tried to use as much light power as possible to measure a high open circuit voltage. In region C the light was kept on and the contact on the ITO was plugged off (outside of the microscope). The solar cell was now operated in open circuit condition and the ITO changed its potential to a higher value. In region D the ITO contact was reconnected again, the solar cell was in short circuit condition and the CPD was similar to that in region A and B. In region D the contact on the Al was plugged off and this time the potential on the Al shifted to a lower level. In region F the illumination was turned off and the measured CPD was the same as for region A, B, and D. This means the measured CPD is the same without light in short and open circuit condition as with light in short circuit condition. Only with one contact floating (open circuit condition) and the light turned on, one can detect a difference in the CPD. Note that due to the drifting, the scan position had to be adjusted several times. Therefore, on a first glance the scan appears uncorrelated, but it was well suitable to take profiles at the different conditions.

In figure 7.8b profiles through region B (black) and region C (red) are shown. B demonstrates the short circuit case and C the open circuit case. One can see that in C the ITO potential rises about 350 mV nearly up to the Al potential. The difference of C and B is the induced photovoltage. It is plotted as blue line and refers to the right y-axis. Most of the open circuit voltage drops at the contact between the ITO and the P3HT and a very small part along the P3HT and the PCBM. In figure 7.8c the profiles for region E and F are shown. E demonstrates the open circuit case with illumination and F demonstrates the open circuit case without illumination. In E the Al potential falls about 350 mV nearly down to the ITO potential. Again, the blue curve shows the difference of E and F which corresponds to the photovoltage and refers to the right y-axis. And again, most of the open circuit voltage drops at the contact between the ITO and the P3HT and a very small part along the P3HT and the PCBM. This shows that the net potential distribution in open circuit condition is independent of which contact is on floating and which contact is

7.1. P3HT/PCBM bilayer solar cells

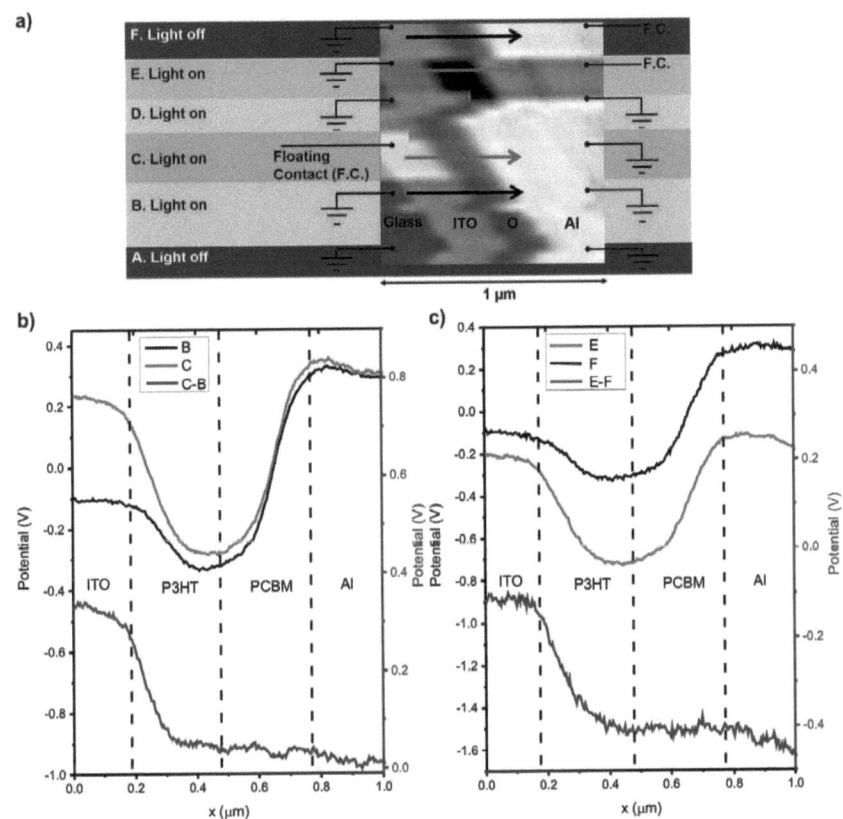

Figure 7.8: a) SKPM scan of a bilayer P3HT/PCBM solar cell (MuBiSpin3_2) under different illumination and electric circuit conditions. The scan was performed from bottom to top and the conditions were changed manually. A: Both contacts grounded, LED off. B: Both contacts grounded, LED on. C: ITO on floating potential, Al grounded, LED on. D: Both contacts grounded, LED on. E: ITO grounded, Al on floating potential, LED on. F: ITO grounded, Al on floating potential, LED off. Profiles of the potentials under the different conditions (in black and red; refer to the left y-axis) and their respective differences (in blue; refers to the right y-axis) are shown in b) and c).

53

7. Potential distribution within P3HT/PCBM BHJ and bilayer solar cells

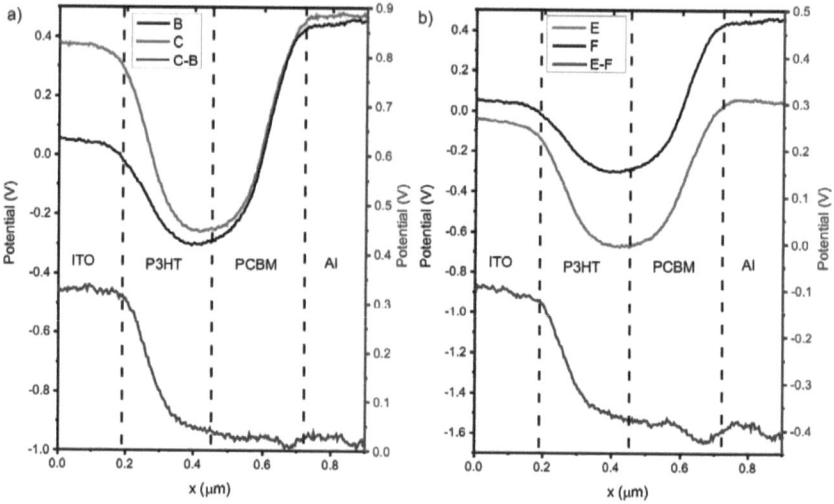

Figure 7.9: SKPM data of a second bilayer P3HT/PCBM solar cell (MuBiSpin3_1). For comparison the data was taken under equal conditions as the measurement in figure 7.8. Profiles of the potentials under the different conditions (in black and red; refer to the left y-axis) and their respective differences (in blue; refers to the right y-axis) are shown. In a) the ITO was on floating potential and Al was kept on ground; in b) Al was on floating potential and ITO was kept on ground.

on ground potential.

The same experiment was also conducted for the second bilayer solar cell MuBiSpin3_1. The measurement was performed exactly in the same way as described before. Figure 7.9a shows the potential distribution with LED illumination in short circuit condition (black curve) and with ITO on ground potential (red curve). The blue line shows the difference of both signals and refers to the right y-axis. In figure 7.9b the potential is shown with ITO on ground potential and Al on floating potential with (red curve) and without illumination (black) curve. Again, the blue line shows the difference of both signals and refers to the right y-axis. In figure 7.9a and b the net potential distribution shows that the open circuit voltage drops mainly on the contact between the ITO and the P3HT and to a small amount within the bilayer.

54

7.2. Conventional P3HT:PCBM BHJ solar cells

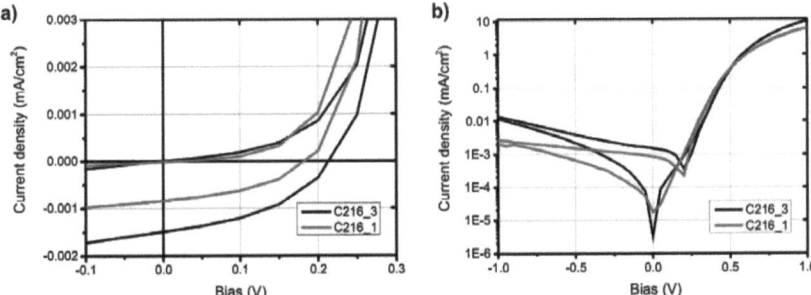

Figure 7.10: IV curves of two P3HT:PCBM BHJ solar cells (C216_1 and C216_3) in the dark and under illumination within the measurement setup in a) linear and b) logarithmic scale.

This confirms the results on the bilayer solar cell MuBiSpin3_2.

7.2. Conventional P3HT:PCBM BHJ solar cells

Conventional P3HT:PCBM BHJ solar cells were the first solar cells investigated within this project. Therefore, the measurement procedure was not as elaborated as in the previous section. Nevertheless, the results were meaningful and reproducible.

7.2.1. IV curves

As described in section 4.5 IV curves of the conventional P3HT:PCBM BHJ solar cells were captured in the measurement setup and the illumination of the solar cells was performed with a fiber where we incoupled the light of a white light source. Figure 7.10 shows the IV curves of the two P3HT:PCBM BHJ solar cells (C216_1 and C216_3) in the dark and under illumination in the measurement setup in a) linear and b) logarithmic scale. The measured open circuit voltage amounts only to about 200 mV in both solar cells which results from the weak illumination. We measured the open circuit voltage outside of the microscope with a bright white light source and a multimeter and obtained values around 500 mV . The reference cells which were made by the same preparation procedure but on the InnovationLab standard layout

7. Potential distribution within P3HT/PCBM BHJ and bilayer solar cells

exhibited open circuit voltages of 550 mV , maximum short circuit currents of about 8 mA/cm² and maximum fill factors of about 70% (compare [88]). Typical efficiencies of the solar cells were about 2.5%.

7.2.2. SKPM data with applied bias voltage in the dark

SKPM scans were captured for different bias voltages similar to the measurements on a bilayer solar cell shown in figure 7.4. From each scan a profile was taken. Figure 7.11 shows the CPD under different bias voltages in a P3HT:PCBM BHJ solar cell (C216_3) for a) -2 V $\rightarrow$ 1.5 V in 0.5 V steps and b) -0 V $\rightarrow$ 0.4 V in 0.1 V steps. The maximum voltage which was possible to apply was 1.5 V. For higher bias voltages the current became to high and the sample heated up too much. c) and d) show the respective net potential distributions. From the net potential distributions one can see that the voltage drops at the interface between ITO and the BHJ and at the interface between the BHJ and Al whereas there is almost no potential drop within the BHJ. For small bias voltages as there are shown in figure 7.11b and d the voltage drops mainly on the contact between the BHJ and the Al.

As a comparison, in figure 7.12a the CPD of a second P3HT:PCBM BHJ solar cell (C216_1) under bias voltages between 0 V $\rightarrow$ 0.4 V in 0.1 V steps is shown. One can see from the net potential distribution shown in figure 7.14b, that for very small bias voltages the voltage drops mainly on the contact between the BHJ and the Al and for higher bias voltages it drops also on the contact between the ITO and the BHJ. There is almost no visible voltage drop in the BHJ. These findings comply with the results of the first BHJ solar cell (C216_3).

7.2.3. SKPM data in open circuit condition

A scan series for different operating modes of the solar cell as already described in detail in subsection 7.1.5 with different illumination and electrical contacting was also executed for the conventional P3HT:PCBM BHJ solar cells. Figure 7.13a shows the SKPM scan under the different conditions. Again, the scan was performed from bottom to top and the conditions were changed manually. In region A both contacts were grounded and the illumination was switched off. In region B both contacts were grounded and the illumination was turned on. In region C the ITO was disconnected and the potential could float, the Al stayed grounded and the illumination was on. In region D both contacts again were grounded and the illumination was still on. In region E the ITO was grounded, the Al was set on floating potential, the illumination was on.

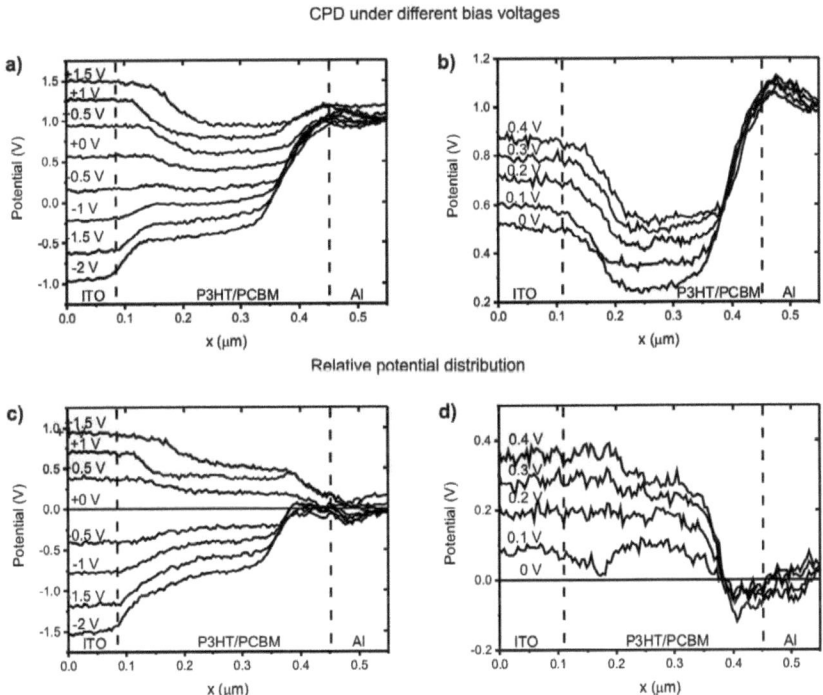

Figure 7.11: CPD under different bias voltages in a P3HT:PCBM BHJ solar cell (C216_3) in the dark for a) -2 V → 1.5 V in 0.5 V steps and b) −0 V → 0.4 V in 0.1 V steps. c) and d) show the respective net potential distributions.

7. Potential distribution within P3HT/PCBM BHJ and bilayer solar cells

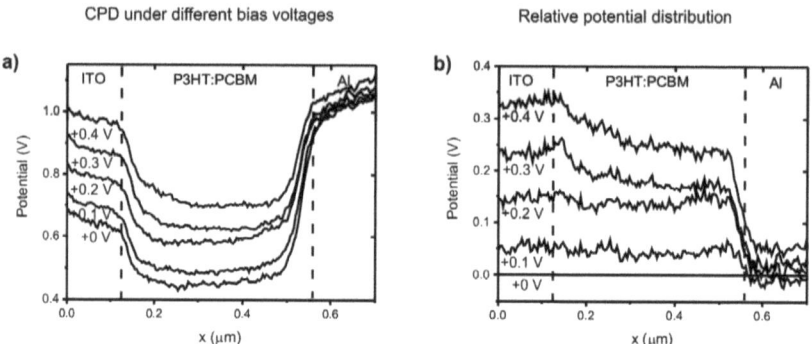

Figure 7.12: a) CPD under different bias voltages in a second P3HT:PCBM BHJ solar cell (C216_1) in the dark for 0 V → 0.4 V in 0.1 V steps. b) shows the respective net potential distributions.

In region F the ITO was grounded, the Al on floating potential and the illumination was turned off. Contrary to the measurements in subsection 7.1.5 here drift played no role as the illumination was very weak and the sample did not heat up measurably. Figure 7.13b shows the CPD profile in region B (black line) and C (red line). This time the ITO potential rises again in direction of the Al potential but only for about 200 mV due to the weak illumination. The blue line shows the difference of both CPDs which corresponds to the induced photovoltage and refers to the right y-axis. In figure 7.13c the CPD profiles of region E and region F are shown. The Al potential under illumination decreases towards the ITO potential but again only for 200 mV. Again, the blue line shows the difference of the two profiles. The net voltage distribution in figure 7.13b and c appear similar such that one can state that the net voltage distribution is independent of which contact was set on floating potential. In all cases the net voltage drop occurs at the contact between the BHJ and the Al.

Figure 7.14 shows the same measurement for the second conventional P3HT:PCBM BHJ solar cell C216_1. In figure 7.14a the CPD profiles under illumination are shown with the Al contact on ground potential. In B (black line) the ITO was set to ground potential, in C (red line) it was on floating potential. The difference between the CPD profiles is drawn in blue and refers to the right y-axis. Again, the main potential drops at the contact between the BHJ and the Al. In figure 7.14b the CPD profiles are shown for the ITO on ground potential and the Al on floating potential. In F (red line) the solar cell is illuminated, in E the illumination was turned off. The blue line shows

7.2. Conventional P3HT:PCBM BHJ solar cells

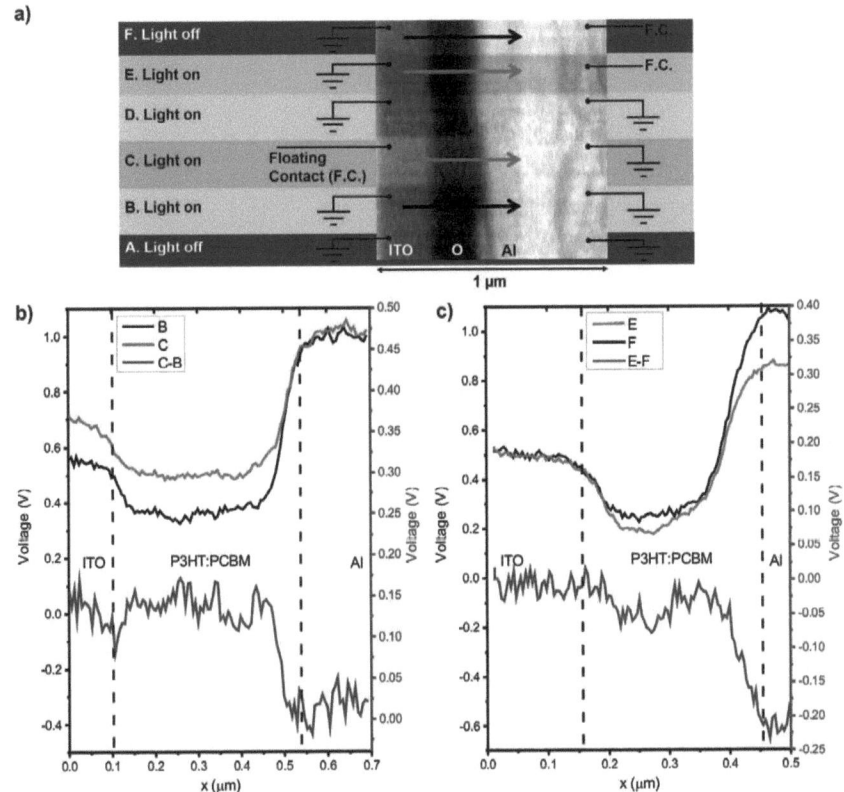

Figure 7.13: a) SKPM scan of a P3HT:PCBM BHJ solar cell (C216_3) under different illumination and electric circuit conditions. The scan was performed from bottom to top and the conditions were changed manually. A: Both contacts grounded, illumination off. B: Both contacts grounded, illumination on. C: ITO on floating potential, Al grounded, illumination on. D: Both contacts grounded, illumination on. E: ITO grounded, Al on floating potential, illumination on. F: ITO grounded, Al on floating potential, illumination off. Profiles of the potentials under the different conditions (in black and red; refer to the left y-axis) and their respective differences (in blue; refers to the right y-axis) are shown in b) and c).

7. *Potential distribution within P3HT/PCBM BHJ and bilayer solar cells*

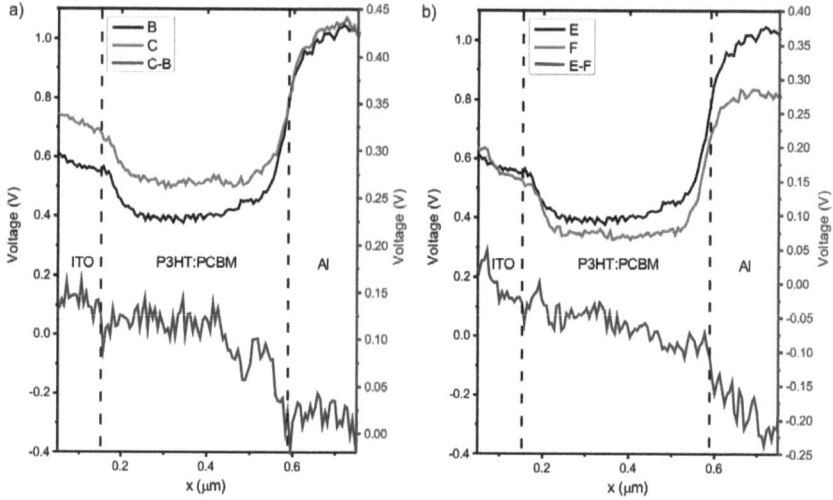

Figure 7.14: SKPM data of a second P3HT:PCBM BHJ solar cell (C216_1) was captured similar to the data displayed in figure 7.13. Profiles of the potentials under the different conditions (in black and red; refer to the left y-axis) and their respective differences (in blue; refers to the right y-axis) are shown. In a) the ITO was on floating potential and Al was kept on ground; in b) Al was on floating potential and ITO was kept on ground.

the difference of the profiles and refers to the right y-axis. This time, the potential drops in the BHJ and at the contact between the BHJ and the Al.

7.3. Inverted P3HT:PCBM BHJ solar cells

As described in chapter 3, in the inverted solar cells the ITO contact is coated with the work function lowering polymer PEIE. The top contact is formed by MoO_3 which has a relatively high work function. Therefore, the bottom contact acts as cathode and the top contact acts as anode. In a consequence, the IV curves are inverted. The following section shows SKPM measurements on these devices.

7.3. Inverted P3HT:PCBM BHJ solar cells

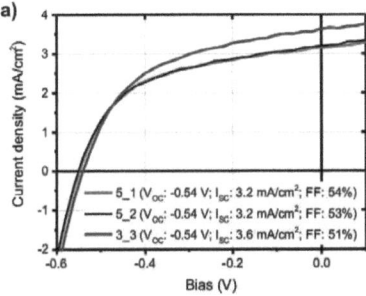
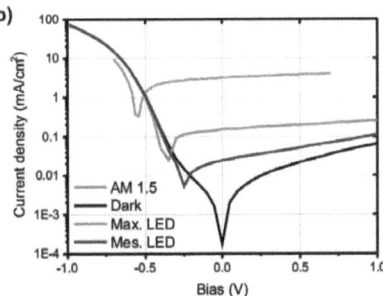

Figure 7.15: a) IV curves of three different inverted P3HT:PCBM BHJ solar cells (Mu5_1, Mu5_2 and Mu3_3) under AM 1.5 illumination. b) IV curves of Mu5_2 under different illumination conditions in logarithmic scale. The dark current is displayed in black, the IV curve under AM 1.5 illumination in orange, the IV curve under maximum LED power in green and the IV curve with illumination power as during the measurements is displayed in blue.

7.3.1. IV curves

Figure 7.15a shows IV curves of three different inverted P3HT:PCBM BHJ solar cells (Mu5_1, Mu5_2 and Mu3_3) under AM 1.5 illumination. Note that these cells are inverted solar cells and therefore with same contacting as for the conventional solar cells, their workpoint lies in the second quadrant. All the solar cells showed an open circuit voltage of -0.54 V, a short circuit current between 3.2 mA/cm^2 and 3.6 mA/cm^2 a fill factor of 51-54%. This results in efficiencies of about 1%. We later found out that our solar simulator was calibrated wrongly and that the light intensity was approximately 40% too low. So the real efficiency should lie about 40% higher. In reference cells with similar processing but on bigger substrates, the efficiencies were in the order of 2.5%. The open circuit voltage was identical in all (functional) cells. In figure 7.15b IV curves of Mu5_2 under different illumination conditions in logarithmic scale are shown. The dark current is displayed in black, the IV curve under AM 1.5 illumination in orange, the IV curve under maximum LED power in green and the IV curve with illumination power as during the measurements is displayed in blue. The curves are similar to the IV curves in figure 5.2b (which belong to MuBi5_1) and show again, that the open circuit voltage in our solar cells depends on the light intensity.

7. Potential distribution within P3HT/PCBM BHJ and bilayer solar cells

7.3.2. SKPM data with applied bias voltage in the dark

For the three inverted P3HT:PCBM BHJ solar cells, the CPD under bias voltage was measured as described in section 7.1.3. In reverse direction, which for inverted solar cells are positive bias voltages, it was possible to apply up to 2 V. In forward direction the sample heated up too much because of too high currents and only 1 V or rather 0.5 V could be applied.

Figure 7.16 shows the CPD profiles of a) Mu5_1, c) Mu5_2 and e) Mu3_3. Figure 7.16b, d and f show the respective net potential distributions. Again, the ITO and the Al are on constant potential and the voltage drops between the contacts. In the inverted solar cells there is no potential drop at the interfaces but the whole voltage drops along the BHJ. This means that the contacts do not form significant charge transport barriers.

7.3.3. SKPM data in open circuit condition

A measurement with different illumination and electrical contacting was also performed for the inverted P3HT:PCBM BHJ solar cells. In figure 7.17 the CPD profiles under illumination are shown with the Al contact on ground potential for a) Mu5_1 and c) Mu5_2. In B (black line) the ITO was set to ground potential and in C (red line) it was on floating potential. The difference between the CPD profiles is drawn in blue and refers to the right y-axis. To minimize the drift due to heating only weak illumination was used. Unfortunately, this lead to low open circuit voltages as can be seen in figure 7.15b. In Mu5_1 the open circuit voltage under this weak illumination amounted only to about 0.1 V. Additionally, the measurement was very noisy such that the distribution of the open circuit voltage is undulated. In Mu5_2 it was at least possible to measure at an open circuit voltage of about 0.2 V which resulted in a slightly better potential profile.

In figure 7.14b the CPD profiles are shown for ITO on ground potential and Al on floating potential. In E (red line) the solar cell is illuminated, in F the illumination was turned off. The blue line shows the difference of the profiles and refers to the right y-axis.

7.4. Discussion

First of all it can be stated that the measurements led to reproducible results. For each type of solar cell at least three different devices were investigated which all showed similar properties. Oftentimes, it was not possible to perform all the different measurements (described at the example of bilayer solar

7.4. Discussion

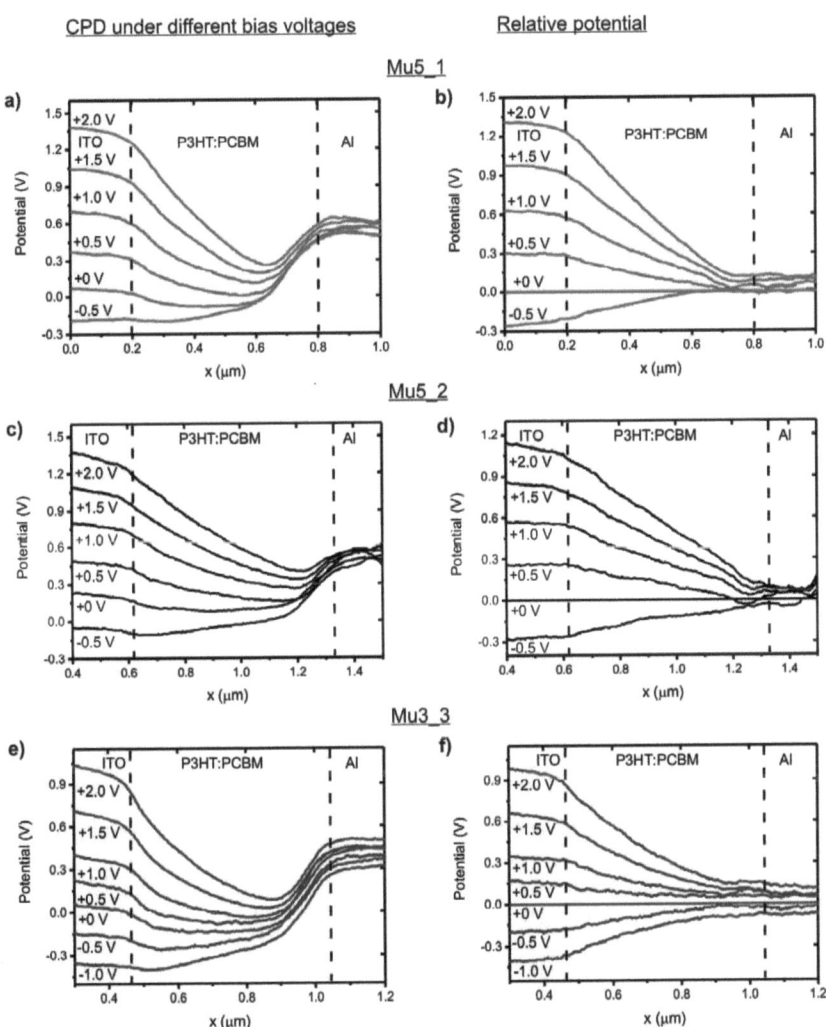

Figure 7.16: CPD under different bias voltages in the inverted P3HT:PCBM BHJ solar cells in the dark. a) Mu5_1, c) Mu5_2 and e) Mu3_3 for $-1\,\text{V} \to 1\,\text{V}$ in $0.5\,\text{V}$ steps. b), d) and f) show their respective relative potential distributions.

7. Potential distribution within P3HT/PCBM BHJ and bilayer solar cells

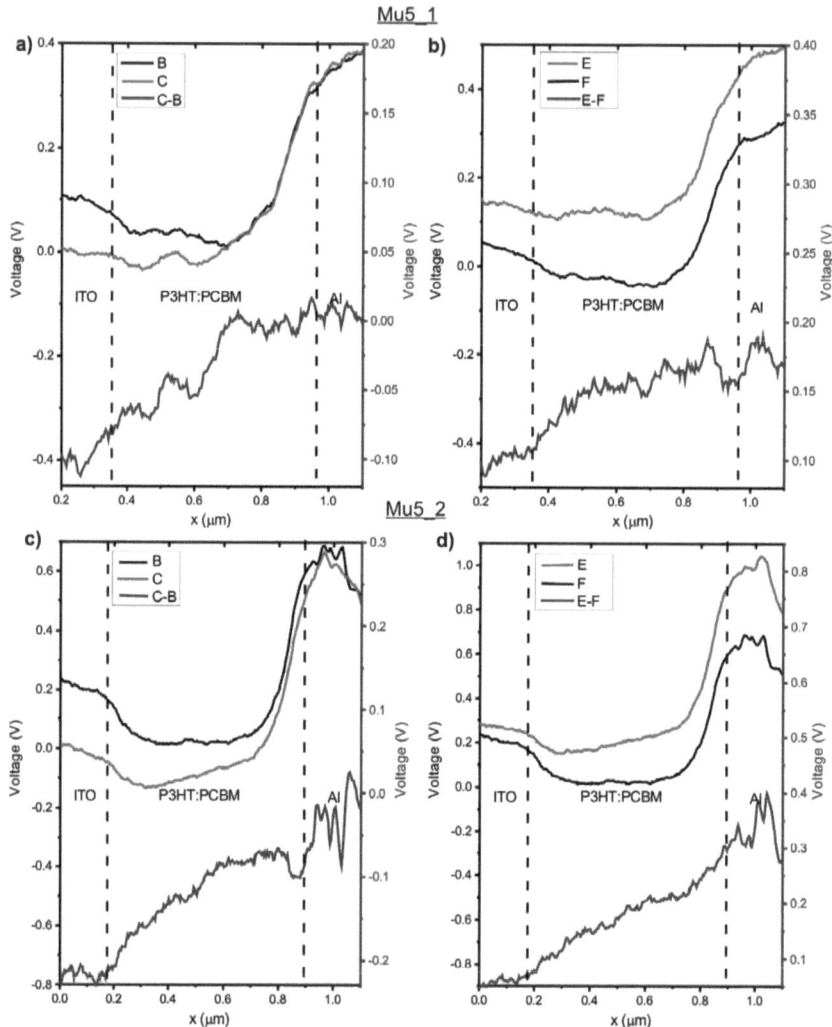

Figure 7.17: SKPM data of the two inverted P3HT:PCBM BHJ solar cells Mu5_1 and Mu5_2 were measured similar to the data displayed in figure 7.13. Profiles of the potentials under the different conditions (in black and red; refer to the left y-axis) and their respective differences (in blue; refers to the right y-axis) are shown. In a) and c) the ITO was on floating potential and Al was kept on ground; in b) and d) Al was on floating potential and ITO was kept on ground.

7.4. Discussion

cell 3_2) on each solar cell. In the preceding sections, a choice of complete measurements was presented. More measurements were executed which were not complete but led to similar results as the above shown measurements. The results on different kinds of solar cells were completely different, although in the case of conventional and inverted BHJ solar cells, only the contacts were changed and the preparation of the BHJ was identical. This indicates that altering of the organic materials due to the FIB milling cannot have a significant influence on the potential distribution.

For bilayer solar cells it was found that an applied bias voltage predominantly drops at the interface between ITO/PEDOT:PSS and P3HT and partly along the BHJ. This shows that PEDOT:PSS coated ITO does not form an ideal contact material for P3HT. On the other hand LiF/Al seems to be a reasonably good contact material for PCBM as no voltage drop was observed at the interface between PCBM and LiF/Al.

In conventional P3HT:PCBM BHJ solar cells for bias voltages higher than 0.5 V the voltage drops on the contact between the ITO/PEDOT:PSS and the BHJ and on the contact between the BHJ and the LiF/Al whereas there is almost no potential drop within the BHJ. For bias voltages lower than 0.5 V the potential dropped mainly on the top contact. This shows that the influence of the injection barriers on the charge transport is voltage dependent. In inverted P3HT:PCBM BHJ solar cells there is no potential drop at the interfaces but the whole voltage drops along the BHJ. So from a charge transport point of view, a PEIE coated ITO bottom contact and a MoO_3/Al top contact form ideal conditions. In figure 7.18 the potential distribution in the different solar cells are summarized. From the measurement on the bilayer, it becomes clear that the ITO/PEDOT:PSS contact forms a transport barrier. So it does not surprise that the conventional BHJ also exhibits a transport barrier at the ITO/PEDOT:PSS contact. In the conventional BHJ, there is also a transport barrier at the top contact, but we know from the measurements on the bilayer cells that PCBM forms a very good contact with LiF/AL. This leads to the conclusion that there is no or only few PCBM on the top contact but that P3HT accumulates on the top. The morphology formation in BHJ materials is a very complex process and has been studied experimentally [153, 154] and by theoretical simulations [155]. The accumulation of P3HT in conventional P3HT:PCBM BHJ solar cells was already observed by other groups [74], e.g. measured by XPS.

Furthermore, an accumulation of P3HT on the top of the BHJ is favorable for the inverted cell layout, which explains that in the inverted solar cells there is no potential drop at the top contact. The charge carrier injection at the top contact of the inverted solar cells can be further improved by a

7. Potential distribution within P3HT/PCBM BHJ and bilayer solar cells

chemical reaction of the MoO_3 with the P3HT [76–78]. Moreover, there is also no potential drop on the bottom contact. Among other things, the morphology formation depends on the surface energy of the bottom contact [154]. The surface energy can be changed by an interfacial dipole layer [156, 157] which in this case is constituted by the PEIE [94, 96, 158]. So it is likely that the PEIE substrate drives the morphology formation to an accumulation of PCBM at the PEIE/BHJ interface, which is optimal for charge transport. Besides it is known that PEIE in general leads to low injection barriers [94, 158, 159]. In any case it can be stated, that in the inverted solar cells the resistance of the organic layer determines the major part of the serial resistance of the solar cell and the contacts play a minor roll. So to further improve this kind of solar cells, the active layer has to be improved, for example by doping of both components or by a change of the morphology to the point of crystalline domains.

From the potential distributions one would expect the inverted BHJ solar cells to be better than the conventional ones. The forward current at 1 V bias voltage in the inverted BHJ solar cell was one order of magnitude higher than in the conventional solar cells, so it can also be seen from the IV curves that the series resistance in the inverted solar cells is lower than in the conventional solar cells. However, the preparation process was not optimized for the inverted BHJ solar cells. Therefore, their efficiencies were comparable.

In all solar cells under short circuit conditions, no change of the potential distribution could be attributed to the illumination.

In open circuit condition, it was possible to image the distribution of the open circuit voltage. Open circuit condition was created by plugging off the bottom or top contact and leaving the other contact on ground potential. The distribution of the open circuit potential was identical for Al floating/ITO grounded and ITO floating/Al grounded. During the measurements of the bilayer solar cells it was tried to have a comparatively powerful illumination. As a disadvantage, the measurement drifted very much, but on the other hand an open circuit voltage of about 0.4 V could be measured. In the measurements of the BHJ solar cells less illumination power was used to avoid drift, but therefore only open circuit voltages of 0.1-0.2 V were built up. As a consequence the obtained distributions of the open circuit voltage are clearer and less noisy for the bilayer solar cells than for the BHJ solar cells. In the bilayer solar cells the major part of the open circuit voltage dropped at the interface between ITO/PEDOT:PSS and P3HT and a minor part dropped along the organic layer. This is identical with the measurements with external applied bias voltage.

In the normal BHJ solar cells the open circuit voltage mainly dropped at

7.4. Discussion

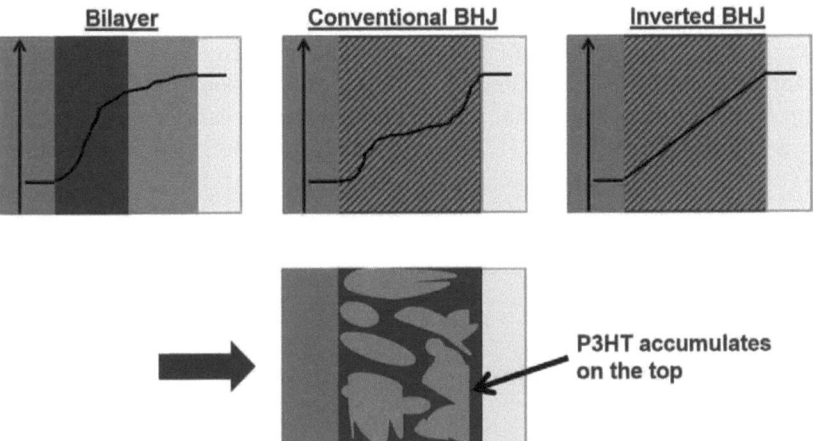

Figure 7.18: Summary of the potential distributions measured in bilayer, conventional BHJ and inverted BHJ P3HT/PCBM solar cells. From these results it can be deduced that the BHJ in our solar cells forms a morphology with accumulation of P3HT on the top which is disadvantageous for conventional BHJ solar cells.

the contact between BHJ and LiF. For external applied bias voltages lower than 0.5 V the potential dropped also on this contact. In the inverted BHJ solar cells, the open circuit voltage distribution was very noisy, but it appears that the open circuit voltage dropped along the organic layer and not on the contacs. Overall, one can state that the distribution of the open circuit voltage within the solar cells was similar as with an externally applied voltage.

As mentioned in chapter 2 there are basically two models to understand the origin of the open circuit voltage in organic solar cells: In the first model, the open circuit voltage is built up by the energy difference between the HOMO level of the donor and the LUMO level of the acceptor components and the contacts have minor influence [81,160]. In the second model, the donor/acceptor interface is needed to separate the exciton but the open circuit voltage is built up by the work function difference of the contact materials [79, 161]. The second model is becoming the more likely one. First of all, because the change of contact materials influences the open circuit voltage dramatically. In the case of our inverted solar cells, the open circuit voltage even changed the sign. Moreover, the presented potential measurements show that the contacts play an important role in organic solar cells. Even in the bilayer solar cell, it was not measured that the open circuit voltage is built up at the organic interface. Therefore, the presented measurements confirm that un upper limit for the

7. Potential distribution within P3HT/PCBM BHJ and bilayer solar cells

generated photo voltage is given by the work function difference of the contact materials. As a comparison: In silicon pn junction solar cells it is well known that the open circuit voltage is limited by the work function difference between the p and the n doped material, which for high doping concentrations nearly corresponds to the band gap [17,61]. But this is only valid if the contacts of the silicon PV device provide ohmic behaviour [17,61].

8. S-shaped solar cells

Recurring difficulties in the processing of (organic) solar cells form so called S-shaped IV curves [162–171]. Under illumination the IV characteristics are not simply shifted to lower current with respect to the IV curve for dark conditions, but exhibits an S-shaped profile. This leads to a greatly reduced fill factor and a reduced open circuit voltage [172].

The origin of S-shaped IV curves is being extensively studied by other scientific work groups in experiments [163–168] as well as in simulations [169–171]. All groups come to the conclusion that transport barriers at the interfaces lead to S-shaped IV curves. This can result from mismatched energy levels at the contacts [165, 166], from interface dipoles [163], from the accumulation of space charges at the interface [164], from reduced surface recombination [169] or imbalanced charge carrier mobilities [170]. In BHJ solar cells S-shaped IV curves can also result from unfavorable phase segregation [168, 169]. As far as it is published in literature, the predicted transport barrier has not yet been directly observed with an imaging method.

During the course of bilayer solar cell investigation it was found that an iodine contamination of the glove box system (e.g. due to Perovskite solar cell fabrication) which was used to fabricate the cells resulted in S-shaped IV characteristics of the solar cells. The iodine contamination was confirmed by XPS measurements [173].[1]

Which on the one hand was a drawback for fabrication of well-functioning devices, is here now used to investigate S-shaped IV curves.

It can be stated that producing solar cells with S-shaped curves is a repro-

---

[1] Batch 1 of the bilayer solar cells was prepared before the contamination at the end of May and this batch did not show S-shaped IV-curves. At the beginning of June, a colleague started the preparation of Perovskite solar cells for which he used material containing iodine. On June 7th, batch 2 was prepared which showed S-shaped IV curves and we know from the XPS measurement [173] that the reference samples were contaminated. After an extensive cleaning procedure of the glove box at the beginning of July, batch 3 was prepared which showed no S-shaped IV curves and iodine was not found in the XPS spectra. On July 22th the glove box was contaminated with the iodine again. Batch 4 and Mu6 (conventional BHJ solar cell) were prepared which again had S-shaped curves. During that time of contamination the preparation of other solar cells [174, 175] did not work either.

8. S-shaped solar cells

ducible effect when contaminating the active layer with iodine, but it would be more sophisticated to address the investigation of S-shaped solar cells by a controlled material or contact modification. Nevertheless it will become clear in this chapter that there is a strong correlation between the IV curves and the potential distribution.

8.1. S-shaped P3HT/PCBM bilayer solar cells

The correlation of certain features in the potential distribution with the characteristics of the IV curve first arised during the investigation of bilayer solar cells. Seven bilayer solar cells in total were investigated by SKPM and all results complied with the results which are going to be discussed in this section.

8.1.1. Results

Figure 8.1 shows the IV curves of two bilayer solar cells N (MuBiSpin3_2) and S (MuBiSpin2_3) in the dark and under LED illumination in a) logarithmic and b) linear scale. As one can see from figure 8.1a sample S shows an open circuit voltage of about $0.3\,\mathrm{V}$ and N shows an open circuit voltage of $0.5\,\mathrm{V}$. The current in forward direction of sample N is about three orders of magnitude higher than the current in sample S although the size of the active region was the same in both cases ($2\,\mathrm{mm} \times 4\,\mathrm{mm}$). In figure 8.1b a zoom into the active region of the solar cells in linear scale is shown. Sample N shows common diode behavior in the dark as well as under illumination. Sample S shows diode behavior only in the dark but under illumination the curve bends. The fill factor of sample S amounts to 13% which is smaller than the fill factor of sample N (53%).

In figure 8.2 the IV curves of altogether three S-shaped bilayer solar cells are shown (MuBiSpin2_1, 2_3 and 4_1) in a) linear and b) logarithmic scale. The solar cells MuBiSpin2_1 and 2_3 show similar curves with a very pronounced anomalous IV curve under illumination. Both have a fill factor of 13% and they were both from batch 2. MuBiSpin4_1 from batch 4 shows less pronounced S-shaped behavior but it is clearly visible (fill factor: 17%). As can be seen from figure 8.2b the current in forward direction of 4_1 is one order of magnitude higher than that of 2_1 and 2_3. The open circuit voltage of 2_1 and 2_3 amounts to about $0.3\,\mathrm{V}$ and the open circuit voltage of 4_1 to about $0.4\,\mathrm{V}$. From figure 8.2b it can also be seen that the current under illumination in backward direction is higher than in forward direction which is

8.1. S-shaped P3HT/PCBM bilayer solar cells

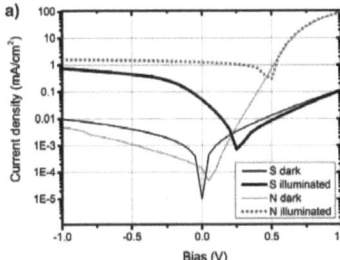 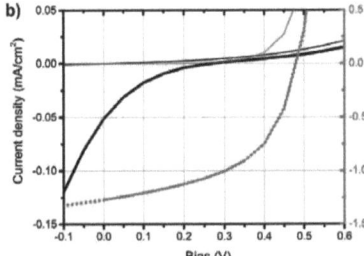

Figure 8.1: From [98]: "IV-curves in the dark and under non standard illumination on a logarithmic scale of sample S (black solid line) and sample N (red dashed line). b) Zoom into the active region of the solar cells. Sample S refers to the left (black) y-axis, sample N to the right (red) y-axis. Sample S exhibits an S-shaped characteristic under illumination".

another property of solar cells with S-shaped IV curves.

In figure 8.3a, c and e the CPD profiles of three different bilayer solar cells (MuBiSpin2_3, 2_2 and 4_1) are shown under different bias voltages between $-1\,\mathrm{V}$ and $1\,\mathrm{V}$ captured in the dark. Figure 8.3b, d and f shows the respective relative potential distributions. In figure 8.3b the relative potential distribution in 2_3 is shown and it is clearly visible that the potential drops exclusively on the top contact between the PCBM and the Al. In figure 8.3d the relative potential distribution of 2_2 is mapped[2]. The resolution of the measurement was not very high due to dust on the cantilever tip. Therefore, the situation is not as clear as in 2_3. But one can definitively state that the voltage drops on the interface between Al and PCBM and it might be that there is also a drop across the PCBM. No potential drop can be seen at the interface between ITO and P3HT and within P3HT. In figure 8.3f the net potential distribution of 4_1 is shown. In 4_1 the major part of the applied voltage drops at the PCBM/Al interface and a minor part at the ITO/P3HT interface and across the organic layers.

Figure 8.4a shows a direct comparison of the net potential distribution in the normal solar cell (MuBiSpin3_2) and the S-shaped solar cells MuBiSpin2_3 and 4_1. In all cases 0.5 V bias voltage was applied. The potential distribution for sample 3_2 is shown in black, the potential distribution for sample 2_3 is shown in red and the potential distribution for sample 4_1 is shown in blue. As the organic layer in 2_3 was smaller than in 3_2 and 4_1, two different

---

[2]2_2 and 2_1 were similar but for 2_1 no SKPM measurements were performed.

71

8. S-shaped solar cells

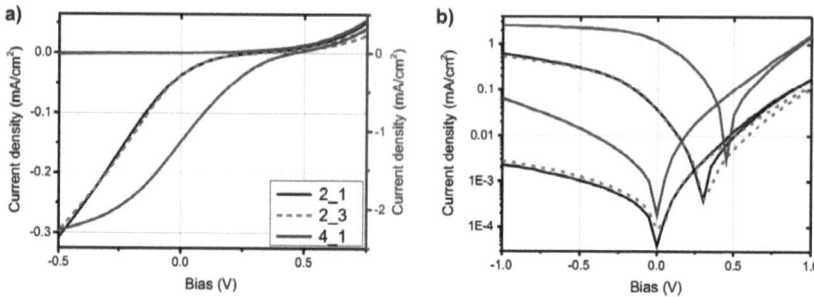

Figure 8.2: IV curves of three different bilayer P3HT/PCBM BHJ solar cells (MuBiSpin2_1, 2_3 and 4_1) in the dark and under LED illumination in a) linear and b) in logarithmic scale. In a) S-shaped characteristics of the solar cells are clearly visible.

x-axis were used to simplify comparison. The lower (black) x-axis refers to sample 3_2 and 4_1, the upper (red) x-axis refers to sample 2_3. In this direct comparison it can be clearly seen that in the normal cell the potential drops at the contact between ITO and P3HT and along the BHJ, but that in sample 2_3 the potential drops exclusively at the contact between the BHJ and the Al. Meanwhile sample 4_1 shows an intermediate state of both cases: The voltage drops at the interface between ITO and P3HT and at the interface between PCBM and Al. Figure 8.4b shows a comparison of the IV curves under LED illumination of the normal solar cell 3_2 and the S-shaped solar cells 2_3 and 4_1. The figure zooms into the active region of the solar cells and three different scale bars are used to display the currents, as the currents were strongly different. The left, black y-axis refers to the black curve which belongs to sample 3_2. The blue curve belongs to sample 4_1 and refers to the right, blue y-axis. The red curve belongs to sample 2_3 and refers to the right, red y-axis. 3_2 shows a standard diode solar cell behavior, 2_3 exhibits a very pronounced S-shaped IV curve. 4_1 shows also an S-shaped IV curve but less pronounced than 2_3.

8.1.2. Discussion

From the IV curves shown in figure 8.1, 8.2 and 8.4 it becomes clear that some of our solar cells exhibited S-shaped IV characteristics. This behavior can be correlated to a temporary iodine contamination of the glove box, further explained above. Compared to the normal bilayer solar cell, the S-shaped

8.1. S-shaped P3HT/PCBM bilayer solar cells

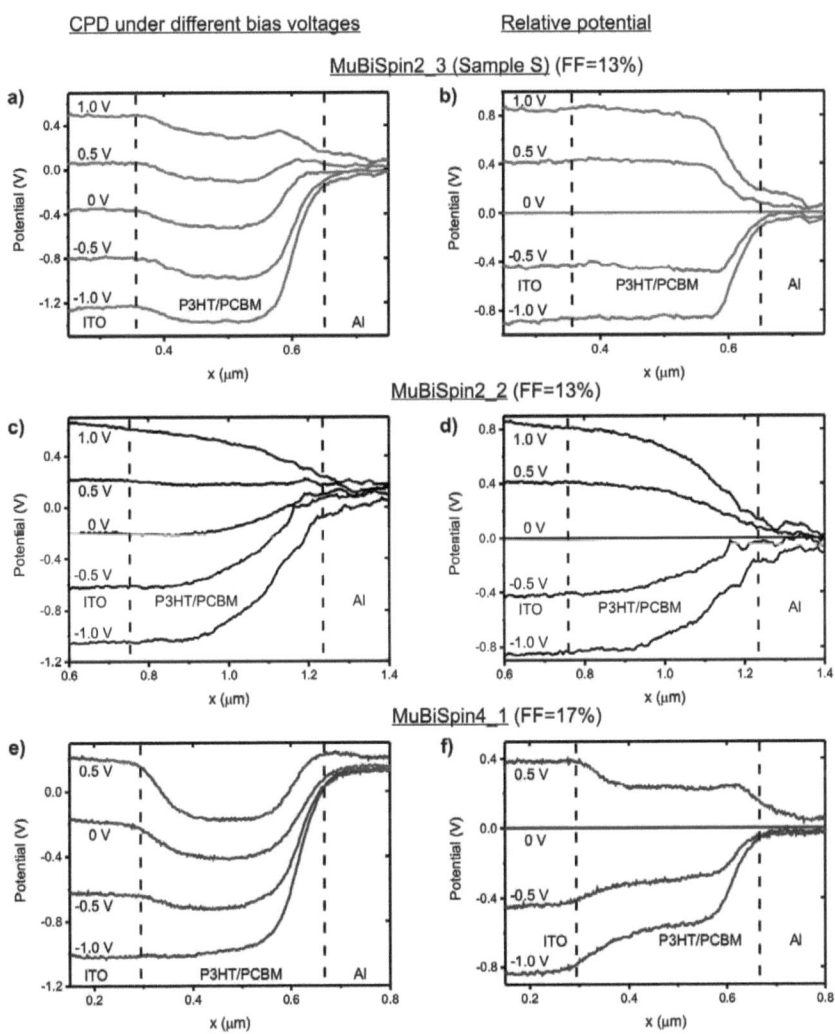

Figure 8.3: CPD under different bias voltages in bilayer P3HT/PCBM solar cells in the dark. a) MuBiSpin2_3, c) MuBiSpin2_2 and e) MuBiSpin4_1 for $-1\,\text{V} \to 1\,\text{V}$ in 0.5 V steps. b), d) and f) show their respective relative potential distributions.

73

8. S-shaped solar cells

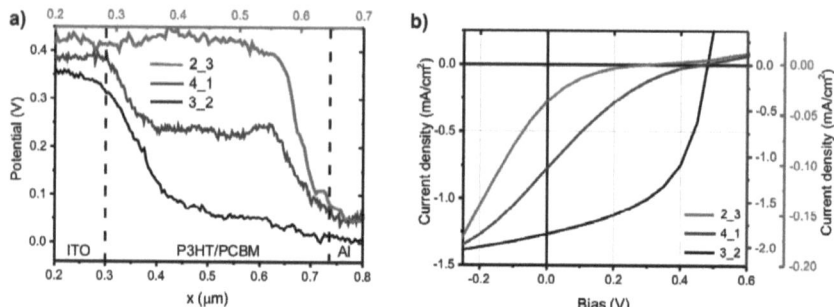

Figure 8.4: a) Comparison of the net potential distributions within the normal solar cell MuBiSpin3_2 and the S-shaped solar cells MuBiSpin2_3 and 4_1 at a bias voltage of 0.5 V in the dark. The potential distribution of 3_2 is displayed in black, the potential distribution of 4_1 in blue and that of 2_3 in red. b) IV curves of the three solar cells under LED illumination. The black curve belongs to cell 3_2 and refers to the left, black y-axis. The blue curve belongs to 4_1 and refers to the right, blue y-axis and the red curve belongs to cell 2_3 and refers to the right, red y-axis.

ones had much lower forward currents. Furthermore, under illumination they showed higher backward current than forward current. The slope of the backward current reminds of a space charge limited current [176, 177] which only occurs in the presence of photoinduced charge carriers [178]. In figure 8.3 the profiles of the CPD under application of different bias voltages are shown. In solar cell 2_3 and 2_2 the voltage dropped exclusively at the interface between PCBM and Al. In solar cell 4_1 a major part of the voltage drop occurred at the PCBM/Al interface and a smaller voltage drop is observed at the ITO/P3HT interface. This is completely different from the potential distribution in normal bilayer solar cells, discussed in chapter 7.1. There, no voltage drop occurred at the PCBM/Al interface. A direct comparison shown in figure 8.4a spotlights the difference and in figure 8.4b the IV curves under illumination are contrasted with the potential profiles. It can be stated that the potential distributions are correlated with the IV curves: the stronger the S-shape behavior is, the higher the potential drop at the PCBM/Al interface is. From the IV curves in the dark in forward direction it can be concluded, that the S-shape solar cells have higher series resistances than the normal solar cells. Calculating the resistances at 0.5 V for each solar cell by dividing 0.5 V through the current of the solar cells at 0.5 V leads to: 0.7 kΩ/cm² for 3_2, 50.1 kΩ/cm² for 2_3 and 5.1 kΩ/cm² for 4_1. It

8.1. S-shaped P3HT/PCBM bilayer solar cells

stands to reason that the higher resistance in the S-shape cells results from an additional transport barrier which is not present in the normal cells. One can think of a very simplified scheme in which the solar cell is described by three different resistances R1, R2 and R3. R1 depicts the resistance between Al and P3HT, R2 is the resistance of the bilayer and R3 the resistance between PCBM and Al. In figure 8.5 this simplified equivalent circuit is shown. As in 3_2 the major part of the voltage drops at the ITO/P3HT interface, in a first order approximation the series resistance of 3_2 ($0.7\,\mathrm{k\Omega/cm^2}$) can be considered as R1 and R2 can be neglected. R3 is different in all solar cells, in 3_2 it is 0, in 4_1 it is R3=$5.1\,\mathrm{k\Omega/cm^2}$-$0.7\,\mathrm{k\Omega/cm^2}$=$4.4\,\mathrm{k\Omega/cm^2}$ and in 2_3 it is R3=$50.1\,\mathrm{k\Omega/cm^2}$ -$0.7\,\mathrm{k\Omega/cm^2}$ =$49.4\,\mathrm{k\Omega/cm^2}$. Note that this is a very simplified model which is only used to explain the static potential distribution for one concrete applied bias voltage. This model does not describe the IV curves of the solar cells! For very discrete interfaces, one would expect a potential step. In reality interfaces are not discrete but rather result in depletion regions. So with SKPM one would expect to measure the potential drop within the size of the depletion zone plus about $20-30\,\mathrm{nm}$ originating from the resolution of SKPM. In figure 8.5 the potential distributions for the equivalent circuit are shown. The steps demonstrate the potential distribution for discrete interfaces. The potential distribution which one would expect to measure within this model is plotted in black for 3_2, in blue for 4_1 and in red for 2_3. The measurement results (compare figure 8.4a) are in very good agreement with this model. However, it has to be mentioned that the height of the potential drop can depend on the region where the profile was taken from, from the zero volt measurement (as this measurement is subtracted from the raw data to gain the net potential distribution) and from the relative position of the zero volt measurement to the measurement under applied bias voltage. As mentioned before, the relative position varies due to drift in the microscope. Therefore, it is hard to make quantitative predictions. But, the scheme shows that the SKPM measurements qualitatively provide results which one would expect if the S-shaped characteristics of the solar cells result from a charge transport barrier at the PCBM/Al interface.

As it was already mentioned, the origin of S-shaped IV curves is being extensively studied by other scientific work groups in experiments [163–168] as well as in theoretical simulations [169–171]. All groups come to the conclusion that transport barriers at the interfaces lead to S-shaped IV curves. This can result from mismatched energy levels at the contacts [165, 166], from interface dipoles [163], from the accumulation of space charges at the interface [164], from reduced surface recombination [169] or imbalanced charge carrier mobilities [170]. So far, from the SKPM measurements it can only be stated that there is

8. S-shaped solar cells

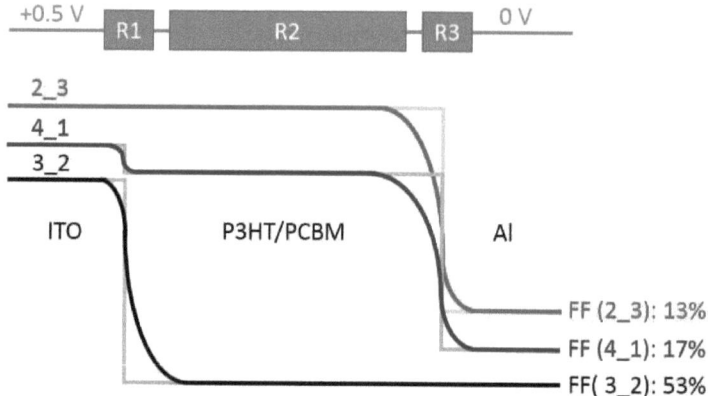

Figure 8.5: Scheme of the potential distribution in the solar cells considering R1=0.7 kΩ/cm^2, R2=0 and R3=0 for 3_2, R3=4.4 kΩ/cm^2 for 4_1 and R3=49.4 kΩ/cm^2 for 2_3.

a transport barrier, but predictions about the properties of this barrier cannot be made. One could think of an energetic barrier in the valence or conduction band but one could also think of reduced mobilities of charge carriers near the interface. Further systematic measurements with and without illumination and with higher bias voltages would be necessary to solve this question. However, it would be wiser to repeat this experiment on systematically prepared S-shaped solar cells as the influence of the iodine contamination may have manifold influences on the solar cell properties [179–181].

Concluding it can be stated that the above shown SKPM measurements for the first time provide a direct mapping of the predicted transport barriers.

8.2. S-shaped P3HT:PCBM BHJ solar cell

S-shaped IV curves were also observed in conventional P3HT:PCBM BHJ solar cells. Similar to the bilayer solar cells, they were fabricated in an iodine contaminated glove box.

8.2.1. Results

In figure 8.6 the IV curve of a conventional P3HT:PCBM BHJ solar cell (Mu6) in the dark (black curve) and under LED illumination (red curve) in the

8.2. S-shaped P3HT:PCBM BHJ solar cell

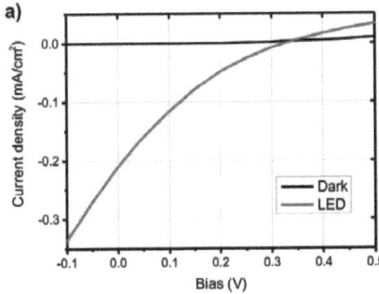 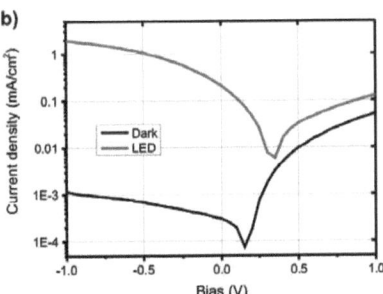

Figure 8.6: IV curve of a conventional P3HT:PCBM BHJ solar cell (Mu6) in the dark and under LED illumination in the measurement setup in a) linear and b) logarithmic scale.

measurement setup are shown in a) linear and b) logarithmic scale. Note that the IV curves were taken in the measurement setup and the laser for readout of the cantilever movement was not swithced off. Therefore, the solar cell shows an open circuit voltage although the LED illumination was turned off. In figure 8.6a a zoom into the active region is shown and the S-shaped characteristics become visible due to the very low fill factor of only 18%. In figure 8.6b it can be seen that the backward current under illumination is higher than the forward current. This is commonly observed in solar cells with S-shaped IV curves.

Figure 8.7a shows the CPD profiles within the solar cell Mu6 with applied bias voltages between −1 V and 1 V in the dark and figure 8.7b shows the corresponding net voltage distribution. It can be seen that the whole voltage drops at the interface between the BHJ and Al.

Figure 8.8 shows a direct comparison of a normal BHJ solar cell (C126_3) and a BHJ solar cell with S-shaped characteristics (Mu6). In figure 8.8a the net potential distribution in both cells at a bias voltage of 0.5 V is shown. It becomes clear that in the normal solar cell the voltage drops at the interface between ITO and BHJ and at the interface between BHJ and Al. In the S-shaped solar cell, the voltage drops exclusively at the top contact, which means at the interface between BHJ and Al. Figure 8.8b shows a direct comparison of the IV curves of both solar cell under illumination in their active region. The IV curve of C126_3 is drawn in black and refers to the left, black y-axis. The IV curve of Mu6 is plotted in red and refers to the right, red y-axis. Note that the illumination of C216_3 was very weak, especially much

8. S-shaped solar cells

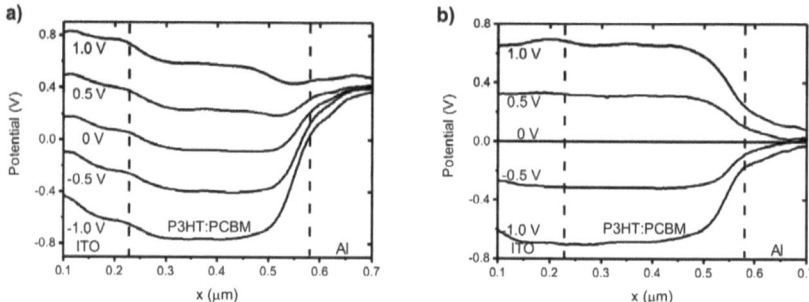

Figure 8.7: a) CPD under different bias voltages in the dark in an S-shaped conventional P3HT:PCBM BHJ solar cell (Mu6) for $-1\,\text{V} \to 1\,\text{V}$ in $0.5\,\text{V}$ steps. b) shows the respective relative potential distributions.

weaker than the illumination of Mu6. Therefore, the current and the open circuit voltage are smaller than they would be under the same illumination power which was used for Mu6. However, from the shape of the curves and from the fill factors (fill factor of C216_3: 46%, fill factor of Mu6: 18%) it becomes clear that C216_3 shows normal diode behavior and Mu6 exhibits S-shaped behavior.

8.2.2. Discussion

The interpretation of the results proceeds similar to the discussion of the previous section. Again, a comparison of the IV curve of the normal solar cell (C216_3), shown in figure 7.10, and the IV curve of the S-shaped solar cell (Mu6), shown in figure 8.6, indicates that the series resistance in the S-shaped solar cell is much higher than in the normal solar cell. However, only one cell which showed S-shaped IV curves was investigated by SKPM. As reproducibility cannot be guaranteed, this results were not published within our paper about: "Understanding S-shaped current-voltage characteristics of organic solar cells: direct measurement of potential distribution by scanning Kelvin probe" [98]. In principle more investigations can be carried out by producing other BHJ solar cell in an iodine contaminated atmosphere.

Calculating the resistances at $0.5\,\text{V}$ for each solar cell by dividing $0.5\,\text{V}$ through the current of the solar cells at $0.5\,\text{V}$ leads to: $1.4\,\text{k}\Omega/\text{cm}^2$ for C216_3 and $51.4\,\text{k}\Omega/\text{cm}^2$ for Mu6. Assuming that the additional series resistance in Mu6 is formed at the BHJ/Al interface, it is reasonable that the whole applied

8.2. S-shaped P3HT:PCBM BHJ solar cell

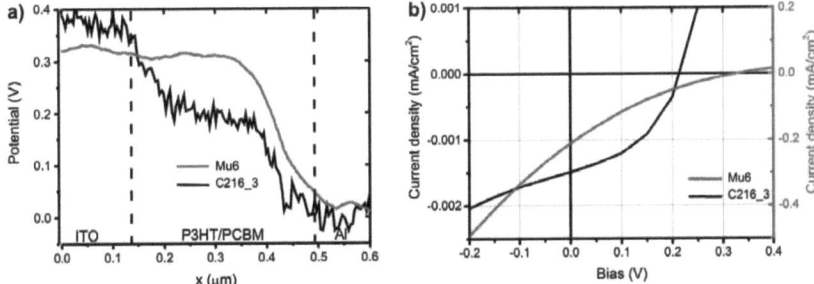

Figure 8.8: a) Comparison of the potential distribution of the two different P3HT:PCBM BHJ solar cells Mu6 (red) and C216_3 (black) in the dark at a bias voltage of 0.5 V. b) IV curves of the solar cells under illumination. The IV curve of C126_3 is drawn in black and refers to the left, black y-axis. The IV curve of Mu6 is plotted in red and refers to the right, red y-axis. C216_3 shows normal solar cell diode behaviour, Mu6 shows a strong S-shaped behaviour. Note that the illumination of C216_3 was very weak, especially much weaker than the illumination of Mu6. Therefore, the current and the open circuit voltage are smaller than they would be under the same illumination power which was used for Mu6.

voltage drops at this interface. This time the solar cell had a BHJ morphology and in contrast to the bilayer solar cells, the P3HT accumulated on the top (compare chapter 7). Therefore, it is likely that the S-shaped IV curves rather result from an imbalanced charge transport than from a change of the energetic interface barrier. Again, the measurements are consistent with literature findings [163–166, 168–171] and for the first time map the potential barrier in a BHJ P3HT:PCBM solar cell with S-shaped IV characteristics.

9. Potential distribution within OLEDs

At the beginning of November 2013 measuremts on cross sections of OLEDs were performed. OLEDs were processed by evaporation of 200 nm N,N´-di-(1-naphthyl)-N,N´-diphenyl-1,1´-biphenyl-4,4´-diamine (NPB) and 200 nm 8-tris-hydroxyquinoline aluminium (Alq_3) on ITO substrates. As top contact a LiF/Al layer was used. The OLEDs were prepared by Christian Weigel (TU Braunschweig). Figure 9.1a shows the CPD profiles in a bilayer OLED (RS0041) at different bias voltages between $-14\,V$ and $5\,V$. The different layers can be clearly distinguished and are marked in the graph. For an applied bias voltage of 0.5 V ITO and Al are approximatly on the same potential so the built in voltage of the device amounts to ca. 0.5 V. In figure 9.1b the zero volt profile is subtracted and the relative potential distribution is obtained. For bias voltages between $-1\,V$ and $1\,V$ the potential drop occurs at the Alq_3/Al interface. For forward bias voltages higher than 1 V the potential drops to a small amount at the Alq_3/Al interface and to a major part along Alq_3. No potential drop occurs along the NPB and at the ITO/NPB interface. For reverse bias voltages between $-1\,V$ and ca. $-5\,V$ the potential drops at the Alq_3/Al interface and along Alq_3. For reverse bias voltages lower than $-5\,V$ the potential drops at the Alq_3/Al interface, along Alq_3 and along NPB.

Stefan Berleb, Wolfgang Brütting et al. performed capacitance-voltage (CV) measurements on such bilayer OLEDs [182, 183]. They used the same bottom contact and the same organic layers but Ca formed the top contact. They found that the capacitance changes at a certain transition voltage which depends on the thicknesses of the organic layers. From the results they deduced a model for the potential distribution within the OLEDs. Interfacial charge Q_{if} between Alq_3 and NPB is held responsible for the predicted potential distribution. In figure 9.2a this model is shown for a bilayer OLED with equal thickness of the organic layers. In this representation the potential distribution is normalized to the flat band condition where the applied bias voltage corresponds to the built in voltage V_{bi} (case (d)). Therefore, the contacts are on the same energy level and so the organic layers are free of an electric field. For an applied forward bias (case (e)) the whole applied voltage drops along Alq_3. For voltages smaller

9. Potential distribution within OLEDs

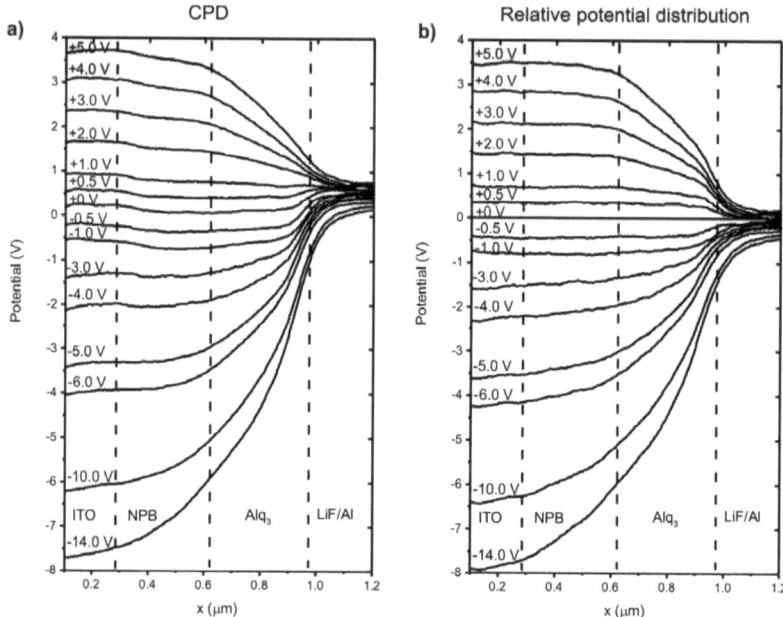

Figure 9.1: a) CPD profiles in a bilayer OLED (RS0041) at different bias voltages between −14 V and 5 V. b) Relative potential distribution in the bilayer OLED at different bias voltages. For bias voltages between −1 V and 1 V the potential drop occurs at the Alq_3/Al interface. For forward bias voltages higher than 1 V the potential drops to a small amount at the Alq_3/Al interface and to a major part along Alq_3. No potential drop occurs along the NPB and at the ITO/NPB interface. For reverse bias voltages between −1 V and ca. −5 V the potential drops at the Alq_3/Al interface and along Alq_3. For reverse bias voltages lower than −5 V the potential drops at the Alq_3/Al interface, along Alq_3 and along NPB.

than V_{bi} but higher than or equal as the transition voltage V_0 (which can be deduced from the CV measurements) the applied voltage drops along Alq$_3$ (case (c) and (b)). For applied voltages higher than V_0 the potential drops to a major part along Alq$_3$ and to a minor part along NPB.

In the presented type of OLED with a 200 nm NPB and a 200 nm Alq$_3$ layer, the transition voltage amounts to ca. -5.0 V [184]. Figure 9.2b shows measured potential profiles at 5.0 V ($>V_{bi}$), 0.5 V ($\approx V_{bi}$), -3.0 V ($V_0 < V < V_{bi}$), -5.0 V ($\approx V_0$) and -10.0 V ($<V_0$). The measured signal is normalized to the built in voltage (≈ 0.5 V) by subtracting the 0.5 V profile from the other profiles, and the y-axis is plotted upside down to facilitate the comparison with figure 9.2a. The obtained potential distributions comply with the model. Only the potential drop at the Alq$_3$/Al interface is not present in the model, but as different contact materials were used a comparison is hardly possible.

Note that already in the raw data shown in figure 9.1a the described behavior can be observed. The representation in figure 9.2b was only used to facilitate the comparison between the model and the measurements.

The very good agreement between the model based on CV measurements and the SKPM results further underline the significance of the presented method as a powerful tool for the investigation of charge transport in organic electronic devices. In particular, these results show that it is possible to distinguish different potential distributions within different organic layers if such is present.

9. Potential distribution within OLEDs

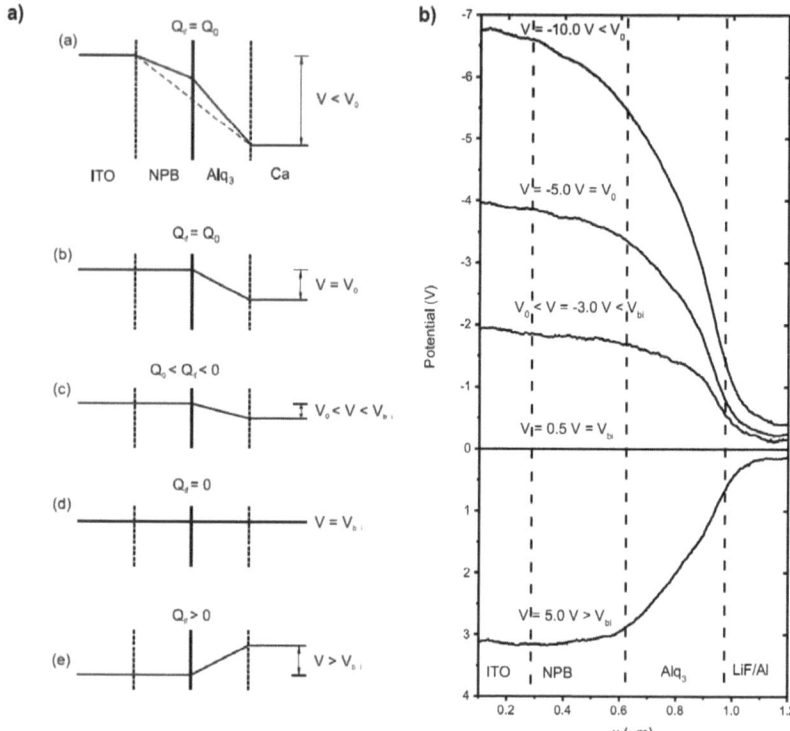

Figure 9.2: a) From [182]: "Spatial dependence of the potential (energy) inside an ITO/NPB/Alq$_3$/Ca device under different bias conditions for equal thickness and dielectric constant of the two organic layers. (a) For large reverse bias the interfacial charge $Q_{if} = Q_0 < 0$ creates a jump of the electrical field, which is equivalent to the change of slope of the potential at the interface. The dashed line indicates the situation without interfacial charge. (b) At $V = V_0$. the flat band condition is reached in NPB. (c) For $V_0 < V < V_{bi}$ NPB stays in the flat band condition, the amount of negative charge at the interface becomes smaller and concomitantly the jump of the electric field is reduced. (d) At $V = V_{bi}$ the flat band condition in both layers is reached and the interfacial charges are fully neutralized. (e) For $V > V_{bi}$ injected positive carriers accumulate at the interface to generate an opposite jump of the electric field". b) Measured potential profiles at 5.0 V ($>V_{bi}$), 0.5 V ($\approx V_{bi}$), -3.0 V ($V_0 < V < V_{bi}$), -5.0 V ($\approx V_0$) and -10.0 V ($<V_0$). The measured signal is normalized to the built in voltage (≈ 0.5 V) by subtracting the 0.5 V profile from the other profiles, and the y-axis is plotted upside down to facilitate the comparison with a).

10. Summary and outlook

In this work an investigation of the potential distribution within organic solar cells was presented. The cross section of the solar cells was laid open by focused ion beam (FIB) milling and afterwards the potential distribution across the cross section was measured by scanning Kelvin probe microscopy (SKPM). A new combined scanning electron microscope (SEM) /FIB cross beam microscope with an implemented scanning probe microscope (SPM) was presented. In this system, FIB milling and further investigation by SPM methods was performed *in-situ*. Furthermore, the system was modified such that the devices could be investigated under operation conditions meaning under illumination and applied bias voltage.

It was found out that the FIB has the potential to dope organic semiconductors and a detailed study on the doping of TIPS-pentacene field effect transistors by FIB was executed. However, the IV curves of organic solar cells were not changed by FIB milling. SKPM measurements were performed on FIB milled, cleaved and microtome cut solar cells and all preparation methods led to similar potential distributions. The interpretation is that the FIB has no significant influence on the measurement results, as the organic layer lies in the shadow of the Al contact during FIB milling.

SKPM measurements on TIPS-pentacene OFETs and silicon solar cells were demonstrated and compared with literature results to proof that SKPM was successfully established in InnovationLab.

Solar cells from Poly(3-hexylthiophen) (P3HT) and [6,6]-phenyl-C61-butyric acid methyl ester (PCBM) were investigated in bilayer, conventional bulk heterojunction (BHJ) and inverted BHJ structure. From contact potential difference (CPD) measurements without applied bias voltage, the relative work functions of the materials could be derived. As expected, Al always showed a lower work function than indium tin oxide (ITO). P3HT and PCBM showed slightly higher work functions than ITO which behaves within the limits of literature results measured by photoelectron spectroscopy and is expected from KP measurements on the materials in InnovationLab.

Under application of bias voltage all different solar cells exhibited different potential distributions. In bilayer P3HT/PCBM solar cells, the applied bias voltage dropped mainly at the Interface between ITO/PEDOT:PSS and P3HT

10. Summary and outlook

and to a minor part within the organic layer. In the conventional BHJ solar cells, the potential dropped at the interface between ITO/PEDOT:PSS and the BHJ and at the interface between the BHJ and LiF/Al, whereas almost no potential dropped within the organic layer. In the inverted BHJ solar cells, there was no potential drop measurable at the contacts but the whole applied voltage dropped within the BHJ. This leads to the conclusion that the inverted BHJ structure is most favorable for P3HT:PCBM BHJ solar cells. Furthermore, it can be concluded, that P3HT accumulates on the top of the BHJ.

Illumination had no influence on the potential distribution of solar cells in short circuit condition. In open circuit condition, the distribution of the open circuit voltage could be measured. Open circuit condition was created by plugging off the bottom or top contact and leaving the other contact on ground potential. The distribution of the open circuit potential was identical for Al floating/ITO grounded and ITO floating/Al grounded. In each solar cell, it had the same distribution as if an external voltage was applied.

Furthermore, bilayer solar cells were fabricated which exhibited S-shaped IV characteristics. SKPM measurements were performed on the cross sections of these devices. It was found that the S-shaped behavior results from an additional transport barrier at the interface between PCBM and LiF/Al which was not present in bilayer solar cells with normal IV characteristics. This finding has been discussed in literature reports, but for the first time, the transport barrier could be directly imaged.

Recently, it was also possible to resolve the potential distribution within bilayer N,N´-di-(1-naphthyl)-N,N´-diphenyl-1,1´-biphenyl-4,4´-diamine (NPB)/8-tris-hydroxyquinoline aluminium (Alq_3) OLEDs. The SKPM measured potential distributions were in accordance with a predictive model based on capacitance voltage measurements from Stefan Berleb, Wolfgang Brütting et al. [182, 183]. These results further underline the capability of the presented measurement method.

All results have shown that the presented method provides meaningful conclusions on the potential distribution and so approaches the understanding of charge transport in organic solar cells. Within this dissertation project, most time was spent to develop, evaluate and optimize the measurement method. Now, after an upgrade and repair of the system, the method is ready to be applied on any electronic device. As it was shown, the method is well suitable to study any kind of interface properties. The influence on the potential distribution by use of different contact materials is one issue which can be explored. The method may also help to investigate the energy alignment in organic materials. Especially the investigation of devices with thick and defined (evaporated) organic layers has the potential of providing new insights.

Furthermore, the measurement setup in principle offers the possibility to use any desired SPM method. For example scanning tunelling microscopy (STM) can be applied to investigate the cross sections. In recent STM measurements by Shih et al. on cleaved cross sections of P3HT:PCBM solar cells, it was possible to resolve the morphology [185]. If STM and SKPM were applied on the same cross section, the potential distribution and the morphology could be directly correlated. Another SPM method which could also be implemented is the so called scanning near field optical microscopy (SNOM). This method could be used to study the local light induced charge carrier generation [186]. It is further possible to use the SEM and investigate the electron beam induced current which provides further information about the local electric field.

The investigation of many different samples with the mentioned methods will significantly help to create a better insight into charge transport in organic materials and will hopefully contribute to an improvement of organic electronic devices.

Bibliography

[1] *International Energy Outlook 2013*. U.S. Energy Information Administration, 2013.

[2] David JC MacKay. *Sustainable Energy -without the hot air*. UIT Cambridge Ltd.

[3] Henning Rodhe. A comparison of the contribution of various gases to the greenhouse effect. *Science*, 248(4960):1217–1219, 1990.

[4] Bert Bolin and Bo R Doos. Greenhouse effect. 1989.

[5] Greg Kopp and Judith L Lean. A new, lower value of total solar irradiance: Evidence and climate significance. *Geophysical Research Letters*, 38(1), 2011.

[6] Thomas B Johansson and Laurie Burnham. *Renewable energy: sources for fuels and electricity*. Island press, 1993.

[7] Martin Pehnt. Dynamic life cycle assessment (LCA) of renewable energy technologies. *Renewable energy*, 31(1):55–71, 2006.

[8] Mark Z Jacobson and Mark A Delucchi. Providing all global energy with wind, water, and solar power, part i: Technologies, energy resources, quantities and areas of infrastructure, and materials. *Energy Policy*, 39(3):1154–1169, 2011.

[9] C Kost, J.N. Mayer, J. Thomson, and N. Hartmann. Stromgestehungskosten Erneuerbare Energien. *Fraunhofer-Institut für Solare Energiesysteme ISE*, 2013.

[10] Erneuerbare-Energien-Gesetz (EEG) 2012.

[11] 24. Subventionsbericht, Bericht der Bundesregierung über die Entwicklung der Finanzhilfen des Bundes und der Steuervergünstigungen für die Jahre 2011 bis 2014.

Bibliography

[12] Michael Valenti. Storing solar energy in salt. *Mechanical Engineering*, 117(6), 1995.

[13] Gregory J Kolb. Economic evaluation of solar-only and hybrid power towers using molten-salt technology. *Solar Energy*, 62(1):51–61, 1998.

[14] MJ Montes, A Abánades, JM Martinez-Val, and M Valdés. Solar multiple optimization for a solar-only thermal power plant, using oil as heat transfer fluid in the parabolic trough collectors. *Solar Energy*, 83(12):2165–2176, 2009.

[15] Albert Einstein. Über einen die Erzeugung und Verwandlung des Lichtes betreffenden heuristischen Gesichtspunkt. *Annalen der Physik*, 322(6):132–148, 1905.

[16] Albert Einstein and Translation into English. Concerning an heuristic point of view toward the emission and transformation of light. *American Journal of Physics*, 33(5):367, 1965.

[17] Peter Wuerfel and Uli Wuerfel. *Physics of Solar Cells: From Basic Principles to Advanced Concepts*. Wiley-VCH, 2009.

[18] Martin A Green. Silicon photovoltaic modules: a brief history of the first 50 years. *Progress in Photovoltaics: Research and Applications*, 13(5):447–455, 2005.

[19] DM Chapin, CS Fuller, and GL Pearson. A new silicon p-n junction photocell for converting solar radiation into electrical power. *Journal of Applied Physics*, 25(5):676–677, 1954.

[20] Adolf Goetzberger and Christopher Hebling. Photovoltaic materials, past, present, future. *Solar energy materials and solar cells*, 62(1):1–19, 2000.

[21] SENSE: Sustainability Evaluation of Solar Energy Systems LCA Analysis. Technical report, University of Stuttgart, 2008.

[22] Hideki Shirakawa, Edwin J Louis, Alan G MacDiarmid, Chwan K Chiang, and Alan J Heeger. Synthesis of electrically conducting organic polymers: halogen derivatives of polyacetylene,(ch) x. *Journal of the Chemical Society, Chemical Communications*, (16):578–580, 1977.

Bibliography

[23] Frederik C Krebs, Thomas Tromholt, and Mikkel Jørgensen. Upscaling of polymer solar cell fabrication using full roll-to-roll processing. *Nanoscale*, 2(6):873–886, 2010.

[24] Claudia N Hoth, Pavel Schilinsky, Stelios A Choulis, and Christoph J Brabec. Printing highly efficient organic solar cells. *Nano letters*, 8(9):2806–2813, 2008.

[25] Sun Hee Lee, Min Hee Choi, Seung Hoon Han, Dong Joon Choo, Jin Jang, and Soon Ki Kwon. High-performance thin-film transistor with 6,13-bis(triisopropylsilylethynyl) pentacene by inkjet printing. *Organic Electronics*, 9(5):721 – 726, 2008.

[26] Chun-Chao Chen, Letian Dou, Rui Zhu, Choong-Heui Chung, Tze-Bin Song, Yue Bing Zheng, Steve Hawks, Gang Li, Paul S Weiss, and Yang Yang. Visibly transparent polymer solar cells produced by solution processing. *Acs Nano*, 6(8):7185–7190, 2012.

[27] Yang Yang. University of California Los Angeles (UCLA). Personal communication.

[28] Martin A. Green, Keith Emery, Yoshihiro Hishikawa, Wilhelm Warta, and Ewan D. Dunlop. Solar cell efficiency tables (version 42). *Progress in Photovoltaics: Research and Applications*, 21(1):827–837, 2013.

[29] Ian A. Howard and Frederic Laquai. Optical probes of charge generation and recombination in bulk heterojunction organic solar cells. *Macromolecular Chemistry and Physics*, 211(19):2063–2070, 2010.

[30] Liam S. C. Pingree, Obadiah G. Reid, and David S. Ginger. Electrical scanning probe microscopy on active organic electronic devices. *Advanced Materials*, 21(1):19–28, 2009.

[31] Evan J Spadafora, Renaud Demadrille, Bernard Ratier, and Benjamin Grévin. Imaging the carrier photogeneration in nanoscale phase segregated organic heterojunctions by kelvin probe force microscopy. *Nano letters*, 10(9):3337–3342, 2010.

[32] Klara Maturova, Martijn Kemerink, Martijn M Wienk, Dimitri SH Charrier, and René AJ Janssen. Scanning kelvin probe microscopy on bulk heterojunction polymer blends. *Advanced Functional Materials*, 19(9):1379–1386, 2009.

Bibliography

[33] J. D. Morris, Timothy L. Atallah, Christopher J. Lombardo, Heungman Park, Ananth Dodabalapur, and X.-Y. Zhu. Mapping electric field distributions in biased organic bulk heterojunctions under illumination by nonlinear optical microscopy. *Applied Physics Letters*, 102(3):033301, 2013.

[34] Kanan P. Puntambekar, Paul V. Pesavento, and C. Daniel Frisbie. Surface potential profiling and contact resistance measurements on operating pentacene thin-film transistors by kelvin probe force microscopy. *Applied Physics Letters*, 83(26):5539–5541, 2003.

[35] Lucile C. Teague, Behrang H. Hamadani, Oana D. Jurchescu, Sankar Subramanian, John E. Anthony, Thomas N. Jackson, Curt A. Richter, David J. Gundlach, and James G. Kushmerick. Surface potential imaging of solution processable acene-based thin film transistors. *Advanced Materials*, 20(23):4513–4516, 2008.

[36] L Burgi, H Sirringhaus, and RH Friend. Noncontact potentiometry of polymer field-effect transistors. *Applied physics letters*, 80(16):2913–2915, 2002.

[37] Edsger C. P. Smits, Simon G. J. Mathijssen, Michael Cölle, Arjan J. G. Mank, Peter A. Bobbert, Paul W. M. Blom, Bert de Boer, and Dago M. de Leeuw. Unified description of potential profiles and electrical transport in unipolar and ambipolar organic field-effect transistors. *Phys. Rev. B*, 76:125202, Sep 2007.

[38] Dieter Wöhrle and Dieter Meissner. Organic solar cells. *Advanced Materials*, 3(3):129–138, 1991.

[39] Michael Scherer, Rebecca Saive, Dominik Daume, Michael Kröger, and Wolfgang Kowalsky. Sample preparation for scanning kelvin probe microscopy studies on cross sections of organic solar cells. *AIP Advances*, 3(9):092134–092134, 2013.

[40] Michael Scherer. Scanning kelvin probe microscopy on organic solar cell cross sections. Diploma thesis, University of Heidelberg, 2012.

[41] http://www.innovationlab.de/en/forschung/spitzencluster-forum-organic-electronics/, 2013.

[42] Franz Schwabl. *Quantenmechanik*. Springer-Verlag Berlin Heidelberg New York, 2002.

Bibliography

[43] Wolfgang Pauli. Über das H-Theorem vom Anwachsen der Entropie vom Standpunkt der neuen Quantenmechanik. *Probleme der modernen Physik, Arnold Sommerfeld zum 60. Geburtstag*, 1928.

[44] Enrico Fermi. Nuclear physics. *University of Chicago Press*, page S. 142, 1950.

[45] H. Bässler. Charge transport in disordered organic photoconductors a monte carlo simulation study. *physica status solidi (b)*, 175(1):15–56, 1993.

[46] Charles E. Mortimer and Ulrich Mueller. Chemie, 2007.

[47] F. Bloch. Über die Quantenmechanik der Elektronen in Kristallgittern. *Zeitschrift für Physik A*, 52:555–600, 1928.

[48] Charles Kittel and Paul McEuen. *Introduction to solid state physics*, volume 7. Wiley New York, 1996.

[49] JL Brédas, J Ph Calbert, DA da Silva Filho, and J Cornil. Organic semiconductors: A theoretical characterization of the basic parameters governing charge transport. *Proceedings of the National Academy of Sciences*, 99(9):5804–5809, 2002.

[50] R. A. Street, J. Zesch, and M. J. Thompson. Effects of doping on transport and deep trapping in hydrogenated amorphous silicon. *Applied Physics Letters*, 43(7):672 –674, oct 1983.

[51] Robert A Street. *Hydrogenated amorphous silicon*. Cambridge University Press, 2005.

[52] C. Klingshirn. *Semiconductor Optics*. Springer Berlin Heidelberg New York, 2007.

[53] Emmy Noether. Invariante Variationsprobleme. *Nachrichten von der Gesellschaft der Wissenschaften zu Göttingen, mathematisch-physikalische Klasse*, 1918:235–257, 1918.

[54] N. D. Ashcroft, N. W.; Mermin. *Solid State Physics*. Thomson, 1976.

[55] A Thellung. Momentum and quasimomentum in the physics of condensed matter. In *Die Kunst of Phonons*, pages 15–32. Springer, 1994.

Bibliography

[56] D. E. Carlson and C. R. Wronski. Amorphous silicon solar cell. *Applied Physics Letters*, 28(11):671–673, 1976.

[57] G. D. Cody, T. Tiedje, B. Abeles, B. Brooks, and Y. Goldstein. Disorder and the optical-absorption edge of hydrogenated amorphous silicon. *Phys. Rev. Lett.*, 47:1480–1483, Nov 1981.

[58] Antoine Kahn, Norbert Koch, and Weiying Gao. Electronic structure and electrical properties of interfaces between metals and π-conjugated molecular films. *Journal of Polymer Science Part B: Polymer Physics*, 41(21):2529–2548, 2003.

[59] W. Schottky. Vereinfachte und erweiterte Theorie der Randschichtgleichrichter. *Zeitschrift fuer Physik*, 118:539–592, 1942.

[60] Nevill Francis Mott and Edward A Davis. *Electronic processes in non-crystalline materials*. Oxford University Press, 2012.

[61] S. M. Sze and K. Kwok. *Physics of Semiconductor Devices*. Wiley and Sons Hoboken New Jersey, 2007.

[62] P. Y Yu and M. Cardona. *Fundamentals of Semiconductors*. Springer Berlin Heidelberg New York, 2003 2. Auflage.

[63] John F. Moulder, William F. Stickle, Peter E. Sobol, and Kenneth D. Bomben. *Handbook of X-ray Photoelectron Spectroscopy*. ULVAC-PHI, Inc., 1995.

[64] Xavier Crispin, Victor Geskin, Annica Crispin, Jerome Cornil, Roberto Lazzaroni, William R Salaneck, and Jean-Luc Brédas. Characterization of the interface dipole at organic/metal interfaces. *Journal of the American Chemical Society*, 124(27):8131–8141, 2002.

[65] A Muoz, N Chetty, and Richard M Martin. Modification of heterojunction band offsets by thin layers at interfaces: Role of the interface dipole. *Physical Review B*, 41(5):2976, 1990.

[66] J Tersoff. Theory of semiconductor heterojunctions: The role of quantum dipoles. *Physical Review B*, 30(8):4874, 1984.

[67] Walter E. Meyerhof. Contact potential difference in silicon crystal rectifiers. *Phys. Rev.*, 71:727–735, May 1947.

Bibliography

[68] FJ Himpsel, G Hollinger, and RA Pollak. Determination of the fermi-level pinning position at si (111) surfaces. *Physical Review B*, 28(12):7014, 1983.

[69] PS Davids, A Saxena, and DL Smith. Nondegenerate continuum model for polymer light-emitting diodes. *Journal of applied physics*, 78(6):4244–4252, 1995.

[70] Carl Tengstedt, Wojciech Osikowicz, William R Salaneck, Ian D Parker, Che-H Hsu, and Mats Fahlman. Fermi-level pinning at conjugated polymer interfaces. *Applied physics letters*, 88(5):053502–053502, 2006.

[71] John Bardeen. Surface states and rectification at a metal semi-conductor contact. *Phys. Rev.*, 71:717–727, May 1947.

[72] JX Tang, CS Lee, ST Lee, and YB Xu. Electronegativity and charge-injection barrier at organic/metal interfaces. *Chemical physics letters*, 396(1):92–96, 2004.

[73] Chongfei Shen and Antoine Kahn. The role of interface states in controlling the electronic structure of Alq3 reactive metal contacts. *Organic electronics*, 2(2):89–95, 2001.

[74] Zheng Xu, Li-Min Chen, Mei-Hsin Chen, Gang Li, and Yang Yang. Energy level alignment of poly(3-hexylthiophene): [6,6]-phenyl c[sub 61] butyric acid methyl ester bulk heterojunction. *Applied Physics Letters*, 95(1):013301, 2009.

[75] Christoph J Brabec, Sean E Shaheen, Christoph Winder, N Serdar Sariciftci, and Patrick Denk. Effect of LiF/metal electrodes on the performance of plastic solar cells. *Applied Physics Letters*, 80(7):1288–1290, 2002.

[76] Yasuo Nakayama, Katsuyuki Morii, Yuuichirou Suzuki, Hiroyuki Machida, Satoshi Kera, Nobuo Ueno, Hiroshi Kitagawa, Yutaka Noguchi, and Hisao Ishii. Origins of Improved Hole-Injection Efficiency by the Deposition of MoO3 on the Polymeric Semiconductor Poly (dioctylfluorene-alt-benzothiadiazole). *Advanced functional materials*, 19(23):3746–3752, 2009.

[77] Sebastian Beck. University of Heidelberg. Personal communication.

[78] Tobias Glaser. University of Heidelberg. Personal communication.

Bibliography

[79] Letian Dou, Jingbi You, Ziruo Hong, Zheng Xu, Gang Li, Robert A. Street, and Yang Yang. 25th anniversary article: A decade of organic/polymeric photovoltaic research. *Advanced Materials*, pages n/a–n/a, 2013.

[80] J Frenkel. On the transformation of light into heat in solids. i. *Physical Review*, 37(1):17, 1931.

[81] C. J. Brabec, A. Cravino, D. Meissner, N. S. Sariciftci, T. Fromherz, M. T. Rispens, L. Sanchez, and J. C. Hummelen. Origin of the open circuit voltage of plastic solar cells. *Advanced Functional Materials*, 11(5):374–380, 2001.

[82] Fukashi Matsumoto, Kazuyuki Moriwaki, Yuko Takao, and Toshinobu Ohno. Synthesis of thienyl analogues of PCBM and investigation of morphology of mixtures in P3HT. *Beilstein Journal of Organic Chemistry*, 4:33, 2008.

[83] Minh Trung Dang, Lionel Hirsch, and Guillaume Wantz. P3HT: PCBM, best seller in polymer photovoltaic research. *Advanced Materials*, 23(31):3597–3602, 2011.

[84] Ze-Lei Guan, Jong Bok Kim, He Wang, Cherno Jaye, Daniel A. Fischer, Yueh-Lin Loo, and Antoine Kahn. Direct determination of the electronic structure of the poly(3-hexylthiophene):phenyl-[6,6]-c61 butyric acid methyl ester blend. *Organic Electronics*, 11(11):1779 – 1785, 2010.

[85] Martin Pfannmoeller, Harald Fluegge, Gerd Benner, Irene Wacker, Christoph Sommer, Michael Hanselmann, Stephan Schmale, Hans Schmidt, Fred A Hamprecht, Torsten Rabe, Rasmus Schroeder, and Wolfgang Kowalsky. Visualizing a homogeneous blend in bulk heterojunction polymer solar cells by analytical electron microscopy. *Nano letters*, 11(8):3099–3107, 2011.

[86] Lars Mueller. Untersuchung des Oberflächenpotentials von organischen Feldeffekttransistoren mittels Raster-Kelvin-Mikroskopie. Bachelor thesis, University of Heidelberg, 2011.

[87] Florian Ullrich. Untersuchungen an organischen Feldeffekttransistoren zum Einfluss von Licht und zur Verwendung von Gemischen aus funktionalisiertem Pentacen und Poly-alpha-Methylstyrol. Bachelor thesis, University of Heidelberg, 2011.

Bibliography

[88] Dominik Daume. C–V–Messungen an P3HT– und PCBM–Einschichtbauteilen sowie P3HT:PCBM–Doppelschicht– und Bulk–Heterojunction–Solarzellen. Master's thesis, University of Heidelberg, 2012.

[89] Christian Mueller. Raster-Kelvin-Mikroskopie an Querschnitten organischer Solarzellen. Master's thesis, University of Heidelberg, 2013.

[90] John E. Anthony, David L. Eaton, and Sean R. Parkin. A road map to stable, soluble, easily crystallized pentacene derivatives. *Organic Letters*, 4(1):15–18, 2002.

[91] Chang Su Kim, Stephanie Lee, Enrique D. Gomez, John E. Anthony, and Yueh-Lin Loo. Solvent-dependent electrical characteristics and stability of organic thin-film transistors with drop cast bis(triisopropylsilylethynyl) pentacene. *Applied Physics Letters*, 93(10):103302 –103302–3, sep 2008.

[92] C.D. Sheraw, T.N. Jackson, D.L. Eaton, and J.E. Anthony. Functionalized pentacene active layer organic thin-film transistors. *Advanced Materials*, 15(23):2009–2011, 2003.

[93] Rebecca Saive, Michael Scherer, Christian Mueller, Dominik Daume, Janusz Schinke, Michael Kroeger, and Wolfgang Kowalsky. Imaging the Electric Potential within Organic Solar Cells. *Advanced Functional Materials*, pages n/a–n/a, 2013.

[94] J. W. Shim, H. Cheun, J. Meyer, C. Fuentes-Hernandez, A. Dindar, Y.H. Zhou, D. K. Hwang, A. Kahn, and B. Kippelen. Polyvinylpyrrolidone-modified indium tin oxide as an electron-collecting electrode for inverted polymer solar cells. *Applied Physics Letters*, 101(7):073303–073303–4, 2012.

[95] Tao Xiong, Fengxia Wang, Xianfeng Qiao, and Dongge Ma. A soluble nonionic surfactant as electron injection material for high-efficiency inverted bottom-emission organic light emitting diodes. *Applied Physics Letters*, 93(12):123310–123310, 2008.

[96] Yinhua Zhou, Canek Fuentes-Hernandez, Jaewon Shim, Jens Meyer, Anthony J. Giordano, Hong Li, Paul Winget, Theodoros Papadopoulos, Hyeunseok Cheun, Jungbae Kim, Mathieu Fenoll, Amir Dindar, Wojciech Haske, Ehsan Najafabadi, Talha M. Khan, Hossein Sojoudi, Stephen Barlow, Samuel Graham, Jean-Luc Brédas, Seth R. Marder, Antoine

Bibliography

Kahn, and Bernard Kippelen. A universal method to produce low-work function electrodes for organic electronics. *Science*, 336(6079):327–332, 2012.

[97] Julian Heusser. Funktionalisierung von Metalloberflächen für organisch elektronische Bauteile. Bachelor thesis, University of Heidelberg, 2012.

[98] Rebecca Saive, Christian Mueller, Janusz Schinke, Robert Lovrincic, and Wolfgang Kowalsky. Understanding S-shaped current-voltage characteristics of organic solar cells: direct measurement of potential distribution by scanning Kelvin probe. *Applied Physics Letters*, 103, 2013.

[99] Alexander L Ayzner, Christopher J Tassone, Sarah H Tolbert, and Benjamin J Schwartz. Reappraising the need for bulk heterojunctions in polymer- fullerene photovoltaics: the role of carrier transport in all-solution-processed P3HT/PCBM bilayer solar cells. *The Journal of Physical Chemistry C*, 113(46):20050–20060, 2009.

[100] Joachim Mayer, Lucille A Giannuzzi, Takeo Kamino, and Joseph Michael. TEM sample preparation and FIB-induced damage. *Mrs Bulletin*, 32(05):400–407, 2007.

[101] Diana Nanova. Technical University of Braunschweig, InnovationLab, University of Heidelberg. Personal communication.

[102] Lord Kelvin. V. contact electricity of metals. *The London, Edinburgh, and Dublin Philosophical Magazine and Journal of Science*, 46(278):82–120, 1898.

[103] D. Cahen and A. Kahn. Electron energetics at surfaces and interfaces: Concepts and experiments. *Advanced Materials*, 15(4):271–277, 2003.

[104] K. Besocke and S. Berger. Piezoelectric driven kelvin probe for contact potential difference studies. *Review of Scientific Instruments*, 47(7):840–842, 1976.

[105] http://www.kelvinprobe.info/index.htm, June 2013.

[106] Masashi Terada, Nobuyuki Nakamura, Yoichi Nakai, Yasuyuki Kanai, Shunsuke Ohtani, Ken-ichiro Komaki, and Yasunori Yamazaki. Observation of an hci-induced nano-dot on an hopg surface with stm and afm. *Nuclear Instruments and Methods in Physics Research Section B: Beam Interactions with Materials and Atoms*, 235(1):452–455, 2005.

Bibliography

[107] Ernst Meyer, Hans Josef Hug, and Roland Bennewitz. *Scanning probe microscopy: the lab on a tip*. Springer, 2003.

[108] http://www.nanoandmore.com/afm-probe-atec-ncpt.html, June 2013.

[109] M. Nonnenmacher, M. P. O'Boyle, and H. K. Wickramasinghe. Kelvin probe force microscopy. *Applied Physics Letters*, 58(25):2921–2923, 1991.

[110] Th Glatzel, S Sadewasser, and M.Ch Lux-Steiner. Amplitude or frequency modulation-detection in kelvin probe force microscopy. *Applied Surface Science*, 210:84 – 89, 2003.

[111] J. M. R. Weaver and David W. Abraham. High resolution atomic force microscopy potentiometry. *Journal of Vacuum Science and Technology B: Microelectronics and Nanometer Structures*, 9(3):1559–1561, 1991.

[112] Wilhelm Melitz, Jian Shen, Andrew C. Kummel, and Sangyeob Lee. Kelvin probe force microscopy and its application. *Surface Science Reports*, 66(1):1 – 27, 2011.

[113] David Coffey. *Expoloring Organic Solar Cells with Scanning Probe Microscopy*. PhD thesis, University of Washington, 2007.

[114] E Strassburg, A Boag, and Y Rosenwaks. Reconstruction of electrostatic force microscopy images. *Review of Scientific instruments*, 76(8):083705–083705, 2005.

[115] Dimitri S. H. Charrier, Martijn Kemerink, Barry E. Smalbrugge, Tjibbe de Vries, and Rene A. J. Janssen. Real versus Measured Surface Potentials in Scanning Kelvin Probe Microscopy. *ACS Nano*, 2(4):622–626, 2008.

[116] George Elias, Thilo Glatzel, Ernst Meyer, Alex Schwarzman, Amir Boag, and Yossi Rosenwaks. The role of the cantilever in kelvin probe force microscopy measurements. *Beilstein journal of nanotechnology*, 2(1):252–260, 2011.

[117] Keith A Brown, Kevin J Satzinger, and Robert M Westervelt. High spatial resolution Kelvin probe force microscopy with coaxial probes. *Nanotechnology*, 23(11):115703, 2012.

[118] Seunghyup Yoo, Benoit Domercq, and Bernard Kippelen. Intensity-dependent equivalent circuit parameters of organic solar cells based on pentacene and C60. *Journal of Applied Physics*, 97(10):103706–103706, 2005.

Bibliography

[119] D. McMullan. Scanning electron microscopy 1928-1965. *Scanning*, 17(3):175–185, 1995.

[120] T E Everhart and R F M Thornley. Wide-band detector for micro-microampere low-energy electron currents. *Journal of Scientific Instruments*, 37(7):246, 1960.

[121] Nan Yao. *Focused ion beam systems: basics and applications*. Cambridge University Press Cambridge, UK, and New York, 2007.

[122] http://microscopy.zeiss.com/microscopy, December 2013.

[123] http://dme-spm.de/remafm.html, June 2013.

[124] Rebecca Saive, Lars Mueller, Eric Mankel, Wolfgang Kowalsky, and Michael Kroeger. Doping of TIPS-pentacene via Focused Ion Beam (FIB) exposure. *Organic Electronics*, 14(6):1570 – 1576, 2013.

[125] T Burchhart, C Zeiner, A Lugstein, C Henkel, and E Bertagnolli. Tuning the electrical performance of ge nanowire mosfets by focused ion beam implantation. *Nanotechnology*, 22(3):035201, 2011.

[126] Yoshiro Hirayama and Hiroshi Okamoto. Electrical properties of ga ion beam implanted gaas epilayer. *Japanese Journal of Applied Physics*, 24(Part 2, No. 12):L965–L967, 1985.

[127] K. Walzer, B. Maennig, M. Pfeiffer, and K. Leo. Highly efficient organic devices based on electrically doped transport layers. *Chemical Reviews*, 107(4):1233–1271, 2007.

[128] Michael Kroeger, Sami Hamwi, Jens Meyer, Thomas Riedl, Wolfgang Kowalsky, and Antoine Kahn. P-type doping of organic wide band gap materials by transition metal oxides: A case-study on molybdenum trioxide. *Organic Electronics*, 10(5):932 – 938, 2009.

[129] Yabing Qi, Swagat K. Mohapatra, Sang Bok Kim, Stephen Barlow, Seth R. Marder, and Antoine Kahn. Solution doping of organic semiconductors using air-stable n-dopants. *Applied Physics Letters*, 100(8):083305, 2012.

[130] Selina Olthof, Shafigh Mehraeen, Swagat K. Mohapatra, Stephen Barlow, Veaceslav Coropceanu, Jean-Luc Brédas, Seth R. Marder, and Antoine Kahn. Ultralow doping in organic semiconductors: Evidence of trap filling. *Phys. Rev. Lett.*, 109:176601, Oct 2012.

Bibliography

[131] RM Langford and AK Petford-Long. Preparation of transmission electron microscopy cross-section specimens using focused ion beam milling. *Journal of Vacuum Science & Technology A: Vacuum, Surfaces, and Films*, 19(5):2186–2193, 2001.

[132] Eduardo Montoya, Sara Bals, Marta D Rossell, Dominique Schryvers, and Gustaaf Van Tendeloo. Evaluation of top, angle, and side cleaned fib samples for tem analysis. *Microscopy Research and Technique*, 70(12):1060–1071, 2007.

[133] MW Phaneuf. FIB for materials science applications-a review. In *Introduction to Focused Ion Beams*, pages 143–172. Springer, 2005.

[134] http://dme-spm.de/afm.html, November 2013.

[135] Adam Tracz, Jeremiasz K. Jeszka, Mark D. Watson, Wojciech Pisula, Klaus Müllen, and Tadeusz Pakula. Uniaxial alignment of the columnar super-structure of a hexa (alkyl) hexa-peri-hexabenzocoronene on untreated glass by simple solution processing. *Journal of the American Chemical Society*, 125(7):1682–1683, 2003.

[136] Stephen Bain. *Kelvin Force Microscopy of Polymer and Small Molecule Thin-Film Transistors*. PhD thesis, University of Southampton, 2011.

[137] R. Shikler, T. Meoded, N. Fried, and Y. Rosenwaks. Potential imaging of operating light-emitting devices using kelvin force microscopy. *Applied Physics Letters*, 74(20):2972–2974, 1999.

[138] Chun-Sheng Jiang, H. R. Moutinho, D. J. Friedman, J. F. Geisz, and M. M. Al-Jassim. Measurement of built-in electrical potential in iii–v solar cells by scanning kelvin probe microscopy. *Journal of Applied Physics*, 93:10035–10040, 2003.

[139] Th. Glatzel, H. Steigert, S. Sadewasser, R. Klenk, and M.Ch. Lux-Steiner. Potential distribution of cu(in,ga)(s,se)2-solar cell cross-sections measured by kelvin probe force microscopy. *Thin Solid Films*, 480–481:177–182, 2005.

[140] Zhenhao Zhang, Michael Hetterich, Uli Lemmer, Michael Powalla, and Hendrik Holscher. Cross sections of operating Cu (In, Ga) Se 2 thin-film solar cells under defined white light illumination analyzed by Kelvin probe force microscopy. *Applied Physics Letters*, 102(2):023903–023903, 2013.

Bibliography

[141] A Schwarzman, E Grunbaum, E Strassburg, E Lepkifker, A Boag, Y Rosenwaks, Th Glatzel, Z Barkay, M Mazzer, and K Barnham. Nanoscale potential distribution across multiquantum well structures: Kelvin probe force microscopy and secondary electron imaging. *Journal of applied physics*, 98(8):084310–084310, 2005.

[142] Kenneth E Bean. Anisotropic etching of silicon. *Electron Devices, IEEE Transactions on*, 25(10):1185–1193, 1978.

[143] H Seidel, L Csepregi, A Heuberger, and H Baumgärtel. Anisotropic etching of crystalline silicon in alkaline solutions i. orientation dependence and behavior of passivation layers. *Journal of the electrochemical society*, 137(11):3612–3626, 1990.

[144] A Breymesser, V Schlosser, D Peiro, C Voz, J Bertomeu, J Andreu, and J Summhammer. Kelvin probe measurements of microcrystalline silicon on a nanometer scale using sfm. *Solar Energy Materials and Solar Cells*, 66:171–177, 2001.

[145] C-S Jiang, HR Moutinho, R Reedy, MM Al-Jassim, and A Blosse. Two-dimensional junction identification in multicrystalline silicon solar cells by scanning kelvin probe force microscopy. *Journal of Applied Physics*, 104(10):104501–104501, 2008.

[146] Y. Park, V. Choong, Y. Gao, B. R. Hsieh, and C. W. Tang. Work function of indium tin oxide transparent conductor measured by photoelectron spectroscopy. *Applied Physics Letters*, 68(19):2699–2701, 1996.

[147] R Schlaf, H Murata, and ZH Kafafi. Work function measurements on indium tin oxide films. *Journal of Electron Spectroscopy and Related Phenomena*, 120(1):149–154, 2001.

[148] Ming-Chung Wu, Yun-Yue Lin, Sharon Chen, Hsueh-Chung Liao, Yi-Jen Wu, Chun-Wei Chen, Yang-Fang Chen, and Wei-Fang Su. Enhancing light absorption and carrier transport of p3ht by doping multi-wall carbon nanotubes. *Chemical Physics Letters*, 468(1):64–68, 2009.

[149] Tobias Jenne. University of Heidelberg. Personal communication.

[150] Jongjin Lee, Jaemin Kong, Heejoo Kim, Sung-Oong Kang, and Kwanghee Lee. Direct observation of internal potential distributions in a bulk heterojunction solar cell. *Applied Physics Letters*, 99(24):243301–243301-4, dec 2011.

Bibliography

[151] Jaemin Kong, Jongjin Lee, Yonkil Jeong, Maengjun Kim, Sung-Oong Kang, and Kwanghee Lee. Biased internal potential distributions in a bulk-heterojunction organic solar cell incorporated with a tiox interlayer. *Applied Physics Letters*, 100:213305, 2012.

[152] Dong Hwan Wang, Do Youb Kim, Kyeong Woo Choi, Jung Hwa Seo, Sang Hyuk Im, Jong Hyeok Park, O Ok Park, and Alan J Heeger. Enhancement of donor–acceptor polymer bulk heterojunction solar cell power conversion efficiencies by addition of au nanoparticles. *Angewandte Chemie*, 123(24):5633–5637, 2011.

[153] Chris McNeill. Morphology of all-polymer solar cells. *Energy Environ. Sci.*, 2011.

[154] David S. Germack, Calvin K. Chan, R. Joseph Kline, Daniel A. Fischer, David J. Gundlach, Michael F. Toney, Lee J. Richter, and Dean M. DeLongchamp. Interfacial segregation in polymer/fullerene blend films for photovoltaic devices. *Macromolecules*, 43(8):3828–3836, 2010.

[155] K Binder. Surface effects on polymer blends and block copolymer melts: theoretical concepts of surface enrichment, surface induced phase separation and ordering. *Acta polymerica*, 46(3):204–225, 1995.

[156] Sergio A. Paniagua, Peter J. Hotchkiss, Simon C. Jones, Seth R. Marder, Anoma Mudalige, F. Saneeha Marrikar, Jeanne E. Pemberton, and Neal R. Armstrong. Phosphonic Acid Modification of Indium Tin Oxide Electrodes: Combined XPS/UPS/Contact Angle Studies. *The Journal of Physical Chemistry C*, 112(21):7809–7817, 2008.

[157] Paul E. Laibinis and George M. Whitesides. .omega.-terminated alkanethiolate monolayers on surfaces of copper, silver, and gold have similar wettabilities. *Journal of the American Chemical Society*, 114(6):1990–1995, 1992.

[158] Hongkyu Kang, Soonil Hong, Jongjin Lee, and Kwanghee Lee. Electrostatically self-assembled nonconjugated polyelectrolytes as an ideal interfacial layer for inverted polymer solar cells. *Advanced Materials*, 24(22):3005–3009, 2012.

[159] Milan Alt. Karlsruhe Institute of Technology (KIT). Personal communication.

Bibliography

[160] Christoph J Brabec. *Organic photovoltaics: concepts and realization*, volume 60. Springer, 2003.

[161] V. D. Mihailetchi, P. W. M. Blom, J. C. Hummelen, and M. T. Rispens. Cathode dependence of the open circuit voltage of polymer:fullerene bulk heterojunction solar cells. *Journal of Applied Physics*, 94(10):6849–6854, 2003.

[162] I.L Eisgruber, J.E Granata, J.R Sites, J Hou, and J Kessler. Blue-photon modification of nonstandard diode barrier in cuinse2 solar cells. *Solar Energy Materials and Solar Cells*, 53(34):367–377, 1998.

[163] Ankit Kumar, Srinivas Sista, and Yang Yang. Dipole induced anomalous s-shape i-v curves in polymer solar cells. *Journal of Applied Physics*, 105(9):094512, 2009.

[164] A. Wagenpfahl, D. Rauh, M. Binder, C. Deibel, and V. Dyakonov. S-shaped current-voltage characteristics of organic solar devices. *Phys. Rev. B*, 82:115306, Sep 2010.

[165] Hui Jin, Markus Tuomikoski, Jussi Hiltunen, Pa?lvi Kopola, Arto Maaninen, and Flavio Pino. Polymer- electrode interfacial effect on photovoltaic performances in poly (3-hexylthiophene): Phenyl-c61-butyric acid methyl ester based solar cells. *The Journal of Physical Chemistry C*, 113(38):16807–16810, 2009.

[166] Wolfgang Tress, Karl Leo, and Moritz Riede. Influence of hole-transport layers and donor materials on open-circuit voltage and shape of i–v curves of organic solar cells. *Advanced Functional Materials*, 21(11):2140–2149, 2011.

[167] Kouki Akaike and Yoshihiro Kubozono. Correlation between energy level alignment and device performance in planar heterojunction organic photovoltaics. *Organic Electronics*, 2012.

[168] Bertrand Tremolet de Villers, Christopher J. Tassone, Sarah H. Tolbert, and Benjamin J. Schwartz. Improving the reproducibility of p3ht:pcbm solar cells by controlling the pcbm/cathode interface. *The Journal of Physical Chemistry C*, 113(44):18978–18982, 2009.

[169] BY Finck and BJ Schwartz. Understanding the origin of the s-curve in conjugated polymer/fullerene photovoltaics from drift-diffusion simulations. *Applied Physics Letters*, 103(5):053306–053306, 2013.

[170] Wolfgang Tress, Annette Petrich, Markus Hummert, Moritz Hein, Karl Leo, and Moritz Riede. Imbalanced mobilities causing s-shaped iv curves in planar heterojunction organic solar cells. *Applied Physics Letters*, 98(6):063301–063301, 2011.

[171] Jenny Nelson, James Kirkpatrick, and P. Ravirajan. Factors limiting the efficiency of molecular photovoltaic devices. *Phys. Rev. B*, 69:035337, Jan 2004.

[172] Christian Uhrich, Rico Schueppel, Annette Petrich, Martin Pfeiffer, Karl Leo, Eduard Brier, Pinar Kilickiran, and Peter Baeuerle. Organic thin-film photovoltaic cells based on oligothiophenes with reduced bandgap. *Advanced Functional Materials*, 17(15):2991–2999, 2007.

[173] Janusz Schinke. Technical University of Braunschweig. Personal communication.

[174] Lars Mueller. University of Heidelberg. Personal communication.

[175] Carsten Leinweber. University of Heidelberg. Personal communication.

[176] Alvin M Goodman and Albert Rose. Double extraction of uniformly generated electron-hole pairs from insulators with noninjecting contacts. *Journal of Applied Physics*, 42(7):2823–2830, 1971.

[177] A. Rose. *Concepts in photocontuctivity and allied problems*. Interscience Publishers, 1963.

[178] V. D. Mihailetchi, J. Wildeman, and P. W. M. Blom. Space-charge limited photocurrent. *Phys. Rev. Lett.*, 94:126602, Apr 2005.

[179] TJ Prosa, MJ Winokur, Jeff Moulton, P Smith, and AJ Heeger. X-ray diffraction studies of iodine-doped poly (3-alkylthiophenes). *Synthetic metals*, 55(1):370–377, 1993.

[180] Kohji Tashiro, Masamichi Kobayashi, Tsuyoshi Kawai, and Katsumi Yoshino. Crystal structural change in poly (3-alkyl thiophene) s induced by iodine doping as studied by an organized combination of x-ray diffraction, infrared/raman spectroscopy and computer simulation techniques. *Polymer*, 38(12):2867–2879, 1997.

[181] Zuliang Zhuo, Fujun Zhang, Jian Wang, Jin Wang, Xiaowei Xu, Zheng Xu, Yongsheng Wang, and Weihua Tang. Efficiency improvement of

Bibliography

polymer solar cells by iodine doping. *Solid-State Electronics*, 63(1):83–88, 2011.

[182] Stefan Berleb, Wolfgang Brütting, and Gernot Paasch. Interfacial charges and electric field distribution in organic hetero-layer light-emitting devices. *Organic Electronics*, 1(1):41–47, 2000.

[183] Wolfgang Brütting, Stefan Berleb, and Anton G Mückl. Device physics of organic light-emitting diodes based on molecular materials. *Organic electronics*, 2(1):1–36, 2001.

[184] Christian Weigel. Technical University of Braunschweig. Personal communication.

[185] Min-Chuan Shih, Bo-Chao Huang, Chih-Cheng Lin, Shao-Sian Li, Hsin-An Chen, Ya-Ping Chiu, and Chun-Wei Chen. Atomic-scale interfacial band mapping across vertically phased-separated polymer/fullerene hybrid solar cells. *Nano letters*, 2013.

[186] C Coluzza, G Di Claudio, S Davy, M Spajer, D Courjon, A Cricenti, R Generosi, G Faini, J Almeida, E Conforto, et al. Photocurrent near-field microscopy of Schottky barriers. *Journal of microscopy*, 194(2-3):401–406, 1999.

A. Appendix

A.1. Notes on the measurement procedure

As the experimental setup contains a completely new SPM some problems can occur which are not described elsewhere. In this section some of the frequently occurring difficulties will be described in order to help future operators.

A.1.1. Mounting of cantilever and sample

Dimensions and tolerances in the microscope are very small. Therefore one has to pay attention on an accurate mode of operation. After cantilever mounting it is important to check that the cantilever is not electrically connected with the cantilever holder but only with the pin for the SKPM signal. To ensure a proper electrical connection between the cantilever and the piano wire which fixes the cantilever, silver glue has to be applied. This has not change by the system upgrade! Furthermore it is important to control that there is no supernatant which can touch any part of the microscope.

The sample has to be mounted with a flat angle and fixed properly. Extreme drifting and strange behavior of approaching and scanning usually can be attributed to mistakes in cantilever or sample mounting. Sometimes, it can be observed with the SEM that the actual scan field is smaller than the set scan field. Then, in most cases the sample is not properly fixed or something on the sample (e.g. the wires which are used for the electric contacting) touches anything in the system, most likely the cantilever holder. Furthermore, if the sample is to thick the scanner does not extend a lot and the pressure between the scanner and the plate at the sample holder is not very high. Therefore, it can happen that the sample plate is not properly moved with the scanner.

If the scanning is not correctly performed and the above mentioned factors can be excluded, one should shutdown the whole system (close software and EM server, AURIGA to standby mode and SPM controller off). If this does not help one should test whether the scanner is loose by carefully touching it with tweezers.

A. Appendix

A.1.2. Strange behavior of SKPM signal

Strange behavior of the SKPM signal usually results from wrong electrical contact of the cantilever. Either the cantilever is connected to ground potential or it is not connected to the pin. In any case, the cantilever holder has to be dismounted from the microscope and the contacting has to be checked.

A.1.3. Drift due to heating

If a high current flows through the device or it is illuminated with high LED power, the sample heats up and therefore rises. In figure 7.8 it can be seen that the measurement drifts when the illumination is turned on. In principle, it is possible to turn the illumination on and then to wait a few minutes until a steady state is reached. Then, the cantilever can be approached and the measurement can be performed.

B. Publications

R. SAIVE, C. MUELLER, J. SCHINKE, R. LOVRINCIC, AND W. KOWALSKY: *Understanding S-shaped current-voltage characteristics of organic solar cells: direct measurement of potential distribution by scanning Kelvin probe.* Applied Physics Letters, 103, 2013.

M. SCHERER, R.SAIVE, D. DAUME, M. KROEGER, AND W. KOWALSKY: *Sample preparation for scanning Kelvin probe microscopy studies on cross sections of organic solar cells.* AIP Advances, 3, 092134, 2013.

R. SAIVE, M. SCHERER, C. MUELLER, D. DAUME, J. SCHINKE, M. KROEGER, AND W. KOWALSKY: *Imaging the Electric Potential within Organic Solar Cells.* Advanced Functional Materials, 2013.

R. SAIVE, L. MUELLER, E. MANKEL, W. KOWALSKY, AND M. KROEGER: *Doping of TIPS-pentacene via Focused Ion Beam (FIB) exposure.* Organic Electronics, 14, 1570-1576, 2013.

A. LAUCHT, S. PÜTZ, T. GÜNTHNER, N. HAUKE, R. SAIVE, S. FRÉDÉRICK, M. BICHLER, M.-C. AMANN, A. W. HOLLEITNER, M. KANIBER, AND J. J. FINLEY.: *A Waveguide-CoupLED On-Chip Single-Photon Source.* Physical Review X 2, 011014, 2012.

A. LAUCHT, T. GÜNTHNER, S. PÜTZ, R. SAIVE, S. FRÉDÉRICK, N. HAUKE, M. BICHLER, M.-C. AMANN, A. W. HOLLEITNER, M. KANIBER, AND J. J. FINLEY.: *Broadband Purcell enhanced emission dynamics of quantum dots in linear photonic crystal waveguides.* Journal of Applied Physics 112, 093520, 2012.

i want morebooks!

Buy your books fast and straightforward online - at one of the world's fastest growing online book stores! Environmentally sound due to Print-on-Demand technologies.

Buy your books online at
www.get-morebooks.com

Kaufen Sie Ihre Bücher schnell und unkompliziert online – auf einer der am schnellsten wachsenden Buchhandelsplattformen weltweit!
Dank Print-On-Demand umwelt- und ressourcenschonend produziert.

Bücher schneller online kaufen
www.morebooks.de

OmniScriptum Marketing DEU GmbH
Heinrich-Böcking-Str. 6-8
D - 66121 Saarbrücken
Telefax: +49 681 93 81 567-9

info@omniscriptum.de
www.omniscriptum.de

Printed by Books on Demand GmbH, Norderstedt / Germany